Cooling Systems Troubleshooting Handbook

Billy C. Langley

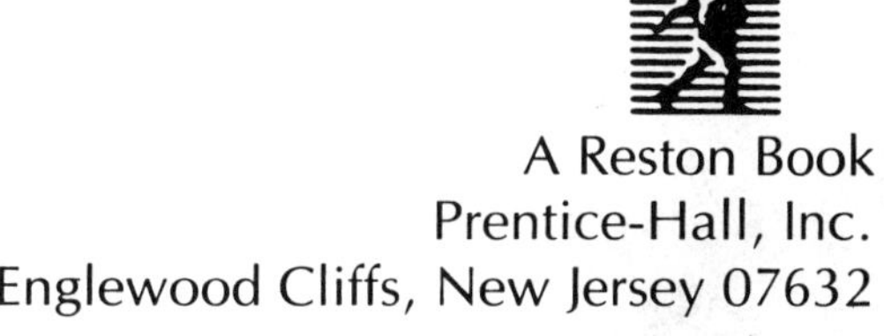

A Reston Book
Prentice-Hall, Inc.
Englewood Cliffs, New Jersey 07632

Library of Congress Cataloging-in-Publication Data

Langley, Billy C.
Cooling systems troubleshooting handbook.
1. Air conditioning — Equipment and supplies — Maintenance and repair. I. Title.
TH7687.7.L37 1986 697.9'3 85-24413
ISBN 0-8359-1036-9

A Reston Book
Published by Prentice-Hall, Inc.
A Division of Simon & Schuster, Inc.
Englewood Cliffs, New Jersey 07632

10 9 8 7 6 5 4 3 2

Printed in the United States of America

CONTENTS

PREFACE

HOW TO USE THIS HANDBOOK

When trouble is experienced with a particular type of equipment, turn to the chart for that type of unit in Chapter 1, Troubleshooting Charts. The action of the unit can then be found in the column labelled "condition." The next column has the "possible causes" that may cause the action encountered. The third column lists the "corrective action" that may be taken to correct the problem. The "reference" column directs the reader to the specific sections in Chapter 2 where the component is introduced and where checkout procedures for the component is given. Specific procedures described in Chapter 4, "Standard Service Procedures," will help the beginner in accomplishing many of the tasks that may be unfamiliar.

Example: We are having trouble with a heat pump unit in the cooling cycle. After checking the unit we find that the compressor runs but there is no cooling. When we refer to the troubleshooting chart, "Heat Pump (Cooling Cycle)" the first cause indicated is defective compressor valves. The corrective action is to replace the valve plate or the compressor. The reference column indicates the specific sections in the text where the proper procedure for this job is discussed. These specific sections are Chapter 2, Sections 2.7 and 2.7.3. A check in Chapter 4, "Standard Service Procedures," will provide additional information on making these repairs.

Also included in this handbook are startup procedures for cooling systems and heat pump systems (Chapter 3), standard service procedures (Chapter 4), electrical troubleshooting procedures and wiring diagrams (Chapter 5), and safety procedures (Chapter 6).

Billy C. Langley

CHAPTER 1

Troubleshooting Charts

AIR CONDITIONING

Condition	Possible Cause	Corrective Action	Reference
Unit will not run.	1. Blown power fuse.	1. Replace fuse and check for cause.	2.3
	2. Thermostat not demanding.	2. Turn on thermostat and set temperature.	2.13; 2.13.1; 2.13.2
	3. Blown transformer fuse.	3. Replace fuse and check for cause.	2.3
	4. Burned out transformer.	4. Replace	2.47; 2.47.1
	5. Faulty wiring or loose connections.	5. Repair wiring or connections.	2.15
Outdoor unit will not run.	1. Blown fuse to outdoor unit.	1. Replace fuse and check for cause.	2.3
	2. Thermostat set too high.	2. Set thermostat.	2.13; 2.13.1; 2.13.2
	3. Burned out contactor coil.	3. Replace coil.	2.10; 2.10.1
	4. Burned contactor contacts.	4. Replace contacts.	2.10; 2.10.2
	5. Compressor overload open.	5. Determine overload and correct.	2.6.5; 2.6.6.; 2.6.7
	6. Off on high-pressure control.	6. See entry "High head pressure."	2.14; 2.14.2

Condition	Possible Cause	Corrective Action	Reference
	7. Off on low-pressure control.	7. See entry "Low suction pressure."	2.14; 2.14.1
	8. Faulty wiring or connections.	8. Repair wiring or connections.	2.15
Compressor will not start.	1. Defective contactor contacts.	1. Replace contacts.	2.10; 2.10.2
	2. Compressor overload open.	2. Determine overload and correct.	2.6.5; 2.6.6; 2.6.7; 2.6.8
	3. Bad starting capacitor.	3. Replace starting capacitor.	2.18; 2.18.1
	4. Bad starting relay.	4. Replace starting relay.	2.19; 2.19.1; 2.19.2; 2.19.3; 2.19.4; 2.19.5
	5. Bad run capacitor.	5. Replace run capacitor.	2.18; 2.18.2
	6. Burned out compressor motor.	6. Repair motor or replace compressor.	2.5; 2.6.1; 2.6.2; 2.6.3; 2.6.4
	7. Stuck compressor.	7. Replace compressor.	2.7; 2.7.1; 2.7.2
Outdoor fan will not start.	1. Faulty wiring or loose connections.	1. Repair wiring or connections.	2.15
	2. Burned out fan motor.	2. Replace fan motor.	2.6; 2.6.1; 2.6.2; 2.6.3; 2.6.4
	3. Bad fan motor bearings.	3. Replace bearings or motor.	2.24; 2.24.2; 2.46; 2.46.5
Compressor hums but will not run.	1. Bad starting capacitor.	1. Replace capacitor.	2.18; 2.18.1
	2. Bad starting relay.	2. Replace starting relay.	2.19; 2.19.1; 2.19.2; 2.19.3; 2.19.4; 2.19.5

Complaint	Possible Cause	Remedy	Reference
	3. Burned out compressor motor.	3. Repair or replace compressor.	2.5; 2.6
	4. Stuck compressor.	4. Replace compressor.	2.7; 2.7.1; 2.7.2; 2.12
	5. Defective contactor contacts.	5. Replace contacts.	2.10; 2.10.2
	6. Three-phase compressor single-phasing.	6. Replace fuse or reset circuit breaker.	2.3
	7. Low voltage.	7. Inform power company.	2.17
Compressor cycling on overload.	1. Bad starting capacitor.	1. Replace starting capacitor.	2.18; 2.18.1
	2. Bad starting relay.	2. Replace starting relay.	2.19; 2.19.1; 2.19.2; 2.19.3; 2.19.4; 2.19.5
	3. Bad running capacitor.	3. Replace running capacitor.	2.18; 2.18.2
	4. Weak overload.	4. Replace overload.	2.6.5; 2.6.6; 2.6.7
	5. Bad contactor contacts.	5. Replace contacts.	2.10; 2.10.2
	6. Low voltage.	6. Inform power company.	2.17
	7. Burned compressor motor.	7. Repair or replace compressor.	2.5; 2.6.1; 2.6.2; 2.6.3; 2.6.4
	8. Refrigerant over-charge.	8. Purge overcharge.	2.26; 2.26.6
	9. Shortage of refrigerant.	9. Repair leak and recharge.	2.27; 2.27.1
	10. High suction pressure.	10. Reduce load or repair compressor.	2.44; 2.44.1; 2.44.2; 2.44.3
	11. Air or non-condensables in system.	11. Purge air or non-condensables.	2.26; 2.26.7

Condition	Possible Cause	Corrective Action	Reference
Compressor off on high-pressure control.	1. Refrigerant over-charge.	1. Purge overcharge	2.26; 2.26.6
	2. Dirty outdoor coil.	2. Clean coil.	2.26; 2.26.2
	3. Slipping outdoor fan belt.	3. Replace or adjust fan belt.	2.26; 2.26.2
	4. Outdoor fan motor not running	4. See previous entry "Outdoor fan motor will not start."	2.5; 2.6.1; 2.6.2; 2.6.3; 2.6.4; 2.46.1; 2.46.2; 2.46.3; 2.46.4; 2.46.5; 2.46.6; 2.46.7
	5. Air or noncondenables in the system	5. Purge air or noncondensables.	2.26; 2.26.7
Compressor cycling or off on low-pressure control.	1. Shortage of refrigerant.	1. Repair leak and recharge.	2.27; 2.27.1
	2. Dirty or defective expansion valve.	2. Clean or replace expansion valve.	2.28; 2.28.1; 2.28.2; 2.28.3; 2.28.4; 2.28.5; 2.28.6
	3. Defective expansion valve power element.	3. Replace power element.	2.28; 2.28.5
	4. Dirty indoor air filters.	4. Clean or replace filters.	2.27; 2.27.2
	5. Dirty indoor coil.	5. Clean coil.	2.27; 2.27.3
	6. Indoor blower belt slipping.	6. Replace or adjust belt.	2.26; 2.26.2
	7. Indoor blower not running.	7. See entry "Indoor blower not running."	2.5; 2.6.1; 2.6.2; 2.6.3; 2.6.4; 2.26
	8. Restriction in refrigeration system.	8. Locate and remove restrictions.	2.30; 2.30.1

Noisy compressor.	1. Loose hold-down bolts.	1. Tighten bolts	2.33; 2.34
	2. Low oil level in compressor.	2. See next entry "Compressor loses oil."	
	3. Defective compressor valves.	3. Replace valves and valve plate.	2.7; 2.7.3
	4. Wrong expansion valve superheat setting.	4. Adjust superheat setting.	2.28
	5. Expansion valve stuck.	5. Repair or replace expansion valve.	2.28; 2.28.2
	6. Poor contact of expansion valve thermal bulb.	6. Improve contact.	2.28; 2.28.1
	7. Overcharge of refrigerant (cap tube system).	7. Purge overcharge.	2.26; 2.26.6
Compressor loses oil.	1. Shortage of refrigerant.	1. Repair leak and recharge system with refrigerant and oil.	2.27; 2.27.1
	2. Low suction pressure.	2. See entry "Low suction pressure."	2.27; 2.27.1, 2.27.2, 2.27.3, 2.27.4, 2.27.5, 2.27.6
	3. Expansion valve stuck open.	3. Repair or replace expansion valve.	2.27; 2.27.2
	4. Restriction in refrigeration system.	4. Locate restriction and remove.	2.30; 2.30.1, 2.30.2, 2.30.3, 2.30.4, 2.30.5, 2.30.6
	5. Refrigerant piping improperly sized.	5. Resize piping.	2.38; 2.38.1, 2.38.2, 2.38.3, 2.39; 2.39.1

Condition	Possible Cause	Corrective Action	Reference
No cooling, but compressor runs continuously.	1. Shortage of refrigerant.	1. Repair leak and recharge system.	2.27; 2.27.1
	2. Defective compressor valves.	2. Replace valves and valve plate or compressor.	2.7; 2.7.3
	3. High suction pressure.	3. See entry "High suction pressure."	
	4. Air or non-condensables in system.	4. Purge air or noncondensables.	2.26; 2.26.7
	5. Wrong expansion valve superheat setting.	5. Adjust superheat setting.	2.28
	6. Dirty or defective expansion valve.	6. Repair or replace expansion valve.	2.28; 2.28.5
	7. Indoor coil dirty.	7. Clean coil.	2.27; 2.27.3
	8. Indoor air filter dirty.	8. Clean or replace filter.	2.27; 2.27.2
	9. Indoor blower belt slipping.	9. Replace or adjust belt.	2.26; 2.26.2
	10. Restriction in refrigerant circuit.	10. Locate and remove restriction.	2.30; 2.30.1, 2.30.2, 2.30.3, 2.30.4, 2.30.5
	11. Dirty indoor coil.	11. Clean coil.	2.26; 2.26.5
Too much cooling, compressor runs continuously.	1. Thermostat setting low.	1. Reset thermostat.	2.13; 2.13.1, 2.13.2
	2. Thermostat in wrong location.	2. Relocate thermostat.	2.13; 2.13.1, 2.13.2
	3. Faulty wiring.	3. Repair wiring.	2.15
Liquid refrigerant flooding compressor (cap tube system).	1. Refrigerant overcharge.	1. Purge overcharge.	2.26; 2.26.6
	2. High head pressure.	2. See entry "High head pressure."	

Problem	Possible Cause	Remedy	Section
	3. Indoor coil dirty.	3. Clean coil.	2.27; 2.27.3
	4. Indoor fan belt slipping.	4. Replace or adjust belt.	2.26; 2.26.2
	5. Indoor air filters dirty.	5. Clean or replace filters.	2.27; 2.27.2
	6. Indoor fan not running.	6. See entry "Indoor blower not running."	2.5; 2.5.1, 2.5.2, 2.5.3, 2.5.4
Liquid refrigerant flooding compressor (expansion valve system).	1. Expansion valve superheat setting wrong.	1. Adjust superheat setting.	2.28
	2. Expansion valve stuck open.	2. Repair or replace expansion valve.	2.28; 2.28.2
	3. Expansion valve thermal bulb loose.	3. Improve contact.	2.28; 2.28.1
	4. Refrigerant overcharge.	4. Purge overcharge.	2.26; 2.26.6
	5. Too cold indoor temperature.	5. Raise thermostat setting.	2.13; 2.13.1, 2.13.2
High head pressure.	1. Refrigerant overcharge.	1. Purge overcharge.	2.26; 2.26.6
	2. High ambient temperature.	2. Provide cooler air to condenser.	2.48
	3. Air or noncondensables in system.	3. Purge air or noncondensables.	2.26; 2.26.7
	4. Excessive loading.	4. Reduce load.	2.42; 2.44; 2.44.1, 2.44.2,
	5. Dirty outdoor coil	5. Clean coil	2.26; 2.26.2
	6. Outdoor fan motor not running.	6. See entry "Outdoor fan motor will not start."	2.5; 2.15
	7. Outdoor fan belt slipping.	7. Replace or adjust fan belt.	2.26; 2.26.2

Condition	Possible Cause	Corrective Action	Reference
Low head pressure.	1. Shortage of refrigerant.	1. Repair leak and recharge system.	2.27; 2.27.1
	2. Defective compressor valves.	2. Replace valves and valve plate or compressor.	2.7; 2.7.3
	3. Low suction	3. See entry "Low suction pressure."	
	4. Low air temperature.	4. Provide warmer air.	2.41; 2.41.1, 2.41.2
High suction pressure.	1. Defective compressor valves.	1. Replace valves and valve plate or compressor.	2.7; 2.7.3
	2. Refrigerant over-charge.	2. Purge overcharge.	2.26; 2.26.6
	3. High head pressure.	3. See above entry, "High head pressure."	
	4. Return air temper-ature high.	4. Provide cooler air.	2.48
	5. Excessive load.	5. Reduce load.	2.42; 2.44; 2.44.1, 2.44.2, 2.44.3
	6. Expansion valve stuck open.	6. Clean or replace valve.	2.28; 2.28.2
Low suction pressure.	1. Refrigerant shortage.	1. Repair leak and recharge.	2.27; 2.27.1
	2. Low return air temperature.	2. Set thermostat higher.	2.13; 2.13.1, 2.13.2
	3. Wrong expansion valve superheat setting.	3. Adjust superheat setting.	2.28
	4. Dirty or defective expansion valve.	4. Clean or replace expansion valve.	2.28; 2.28.1, 2.28; 2.28.1, 2.28.2, 2.28.3, 2.28.4, 2.28.5, 2.28.6

	5. Defective expansion valve power element.	5. Replace power element.	2.28.5
	6. Indoor blower belt slipping.	6. Replace or adjust belt.	2.26; 2.26.2
	7. Indoor blower not running.	7. See next entry "Indoor blower not running."	
	8. Restriction in refrigerant circuit.	8. Locate restriction and remove.	2.30; 2.30.1, 2.30.2, 2.30.3, 2.30.4, 2.30.5
	9. Indoor air filters dirty.	9. Clean coil. filter.	2.27; 2.27.3
	10. Outdoor coil dirty.	10. Clean coil.	2.27; 2.27.3
	11. Indoor coil icing.	11. See entry "Indoor coil icing."	
	12. Restricted cap tube.	12. Replace cap tube.	2.30; 2.30.4
Indoor blower not running.	1. Blown fuse.	1. Replace fuse and correct cause.	2.4
	2. Indoor fan relay defective.	2. Replace relay.	
	3. Burned indoor fan motor.	3. Replace fan motor.	2.46; 2.46.1, 2.46.2, 2.46.3, 2.46.4, 2.46.5, 2.46.6
	4. Broken belt.	4. Replace belt.	2.26; 2.26.1
	5. Faulty wiring or connections.	5. Repair wiring or connections.	2.15
Indoor coil icing.	1. Shortage of refrigerant.	1. Repair leak and recharge system.	2.27; 2.27.1
	2. Low suction pressure.	2. See entry "Low suction pressure."	
	3. Low return air temperature.	3. Raise thermostat setting.	2.26; 2.26.1, 2.26.2
	4. Indoor blower not running.	4. See entry "Indoor blower not running."	

Condition	Possible Cause	Corrective Action	Reference
	5. Indoor blower belt slipping.	5. Replace or adjust belt.	2.26; 2.26.2
	6. Restriction in refrigerant system.	6. Locate and remove restriction.	2.30; 2.30.1; 2.30.2, 2.30.3, 2.30.4, 2.30.5
	7. Indoor air filter dirty.	7. Clean or replace filter.	2.27; 2.27.2
	8. Dirty indoor coil.	8. Clean coil.	2.27; 2.27.3
	9. Dirty or defective expansion valve.	9. Clean or replace expansion valve.	2.27; 2.27.5
High operating costs.	1. Defective compressor valves.	1. Replace valves and valve plate or compressor.	2.7; 2.7.3
	2. Shortage of refrigerant.	2. Repair leak and recharge system.	2.27; 2.27.1
	3. Refrigerant over-charge.	3. Purge overcharge.	2.26; 2.26.6
	4. Dirty outdoor coil.	4. Clean coil.	2.26; 2.26.2
	5. Dirty indoor coil.	5. Clean coil.	2.27; 2.27.3
	6. Dirty indoor air filter	6. Clean or replace filter	2.27; 2.27.2
	7. High head pressure.	7. See entry "High head pressure."	
	8. Thermostat in wrong location.	8. Relocate thermostat.	2.13; 2.13.1, 2.13.2, 2.13.3
	9. Air ducts not insulated.	9. Insulate ducts.	2.52
	10. Unit too small.	10. Install proper size unit.	2.53
	11. Indoor or outdoor fan belt slipping.	11. Replace or adjust belt.	2.26; 2.26.2
	12. Outdoor thermostat set too high.	12. Adjust thermostat.	2.13; 2.13.5

HEAT PUMP (COOLING CYCLE)

Condition	Possible Cause	Corrective Action	Reference
No cooling, but compressor runs continuously.	1. Defective compressor valves.	1. Replace valves and valve plate or compressor.	2.7; 2.7.3
	2. Shortage of refrigerant.	2. Repair leak and recharge system.	2.27; 2.27.1
	3. Defective reversing valve.	3. Replace reversing valve.	2.60
	4. Air or non-condensables in system.	4. Purge non-condensables.	2.26; 2.26.7
	5. Wrong superheat setting on indoor expansion valve.	5. Adjust superheat setting.	2.28
	6. Loose thermal bulb on indoor expansion valve.	6. Tighten thermal bulb.	2.28; 2.28.1
	7. Dirty indoor coil.	7. Clean coil.	2.27; 2.27.3
	8. Dirty indoor air filter.	8. Clean or replace filter.	2.27; 2.27.2
	9. Indoor blower belt slipping.	9. Replace or adjust belt.	2.26; 2.26.2
	10. Restriction in refrigerant system.	10. Locate and remove restriction.	2.29; 2.29.1, 2.29.2, 2.29.3, 2.29.4, 2.29.5
Too much cooling; compressor runs continuously.	1. Faulty wiring.	1. Repair wiring.	2.15
	2. Faulty thermostat.	2. Replace thermostat.	2.26; 2.26.1, 2.26.2
	3. Wrong thermostat location.	3. Relocate thermostat.	

Liquid refrigerant flooding compressor (TXV system).	1. Wrong superheat setting on indoor expansion valve.	1. Adjust superheat.	2.28
	2. Loose thermal bulb on indoor expansion valve.	2. Tighten thermal bulb.	2.28; 2.28.1
	3. Faulty indoor expansion valve.	3. Replace expansion valve.	2.28; 2.28.5
	4. Defective indoor check valve.	4. Replace check valve.	2.59
	5. Refrigerant over-charge.	5. Purge overcharge.	2.26; 2.26.6
Liquid refrigerant flooding com-pressor (capillary tube system).	1. Refrigerant over-charge.	1. Purge overcharge.	2.26; 2.26.6
	2. High Head pressure	2. See Entry "High head pressure."	
	3. Dirty indoor filter.	3. Clean or replace filter.	2.27; 2.27.2
	4. Dirty indoor coil.	4. Clean coil.	2.27; 2.27.3
	5. Indoor blower belt	5. Replace or adjust slipping belt.	2.26; 2.26.2
	6. Indoor check valve defective.	6. Replace check valve.	2.59

HEAT PUMP (HEATING OR COOLING CYCLE)

Condition	Possible Cause	Corrective Action	Reference
Compressor hums but will not start.	1. Faulty fuse.	1. Replace fuse and correct cause.	2.3
	2. Faulty wiring.	2. Repair wiring.	2.15
	3. Loose electrical terminals.	3. Repair loose terminals.	2.15

	4. Compressor overloaded.	4. Locate and remove overload.	2.53
	5. Faulty starting capacitor.	5. Replace capacitor.	2.18; 2.18.1
	6. Faulty starting relay.	6. Replace starting relay.	2.19; 2.19.1, 2.19.2, 2.19.3, 2.19.4, 2.19.5, 2.20
	7. Burned compressor motor.	7. Replace compressor.	2.4; 2.5.1; 2.5.2, 2.5.3., 2.5.4
	8. Defective compressor bearings.	8. Replace bearings or compressor.	2.6; 2.6.1, 2.6.2
	9. Stuck compressor.	9. Replace compressor.	2.6; 2.6.1, 2.6.2
Compressor cycling on overload.	1. Low voltage.	1. Determine reason and repair.	2.17
	2. Loose electrical terminals.	2. Repair terminals.	2.15
	3. Single-phasing of three-phase power.	3. Replace fuse or repair wiring; or notify power company.	2.3; 2.15; 2.17
	4. Defective contactor contacts.	4. Replace contacts or contactor.	2.5; 2.5.2
	5. Defective compressor overload.	5. Replace overload.	2.6.5, 2.6.6, 2.6.7, 2.6.8, 2.6.9, 2.6.10, 2.6.10, 2.6.11, 2.6.12
	6. Compressor overloaded	6. Locate and remove overload.	2.53
	7. Defective start capacitor.	7. Replace capacitor.	2.18; 2.18.1
	8. Defective run capacitor.	8. Replace capacitor.	2.18; 2.18.2

Condition	Possible Cause	Corrective Action	Reference
	9. Defective starting relay.	9. Replace starting relay.	2.19; 2.19.1, 2.19.2, 2.19.3, 2.19.4, 2.19.5; 2.20
	10. Refrigerant overcharge.	10. Purge overcharge.	2.24; 2.24.6
	11. Defective compressor bearings.	11. Replace bearings or compressor.	2.7; 2.7.1, 2.7.2
	12. Air or noncondensables in system (high head pressure).	12. Purge noncondensables from system.	2.24; 2.24.6
	13. Defective reversing valve.	13. Replace reversing valve.	2.60
Compressor off on high pressure control.	1. Refrigerant overcharge.	1. Purge overcharge.	2.24; 2.24.6
	2. Control out of adjustment.	2. Adjust control.	2.14; 2.14.2
	3. Defective indoor fan motor.	3. Repair or replace motor.	2.5; 2.6.1, 2.6.2, 2.6.3, 2.6.4; 2.26; 2.26.1, 2.26.2, 2.26.3, 2.26.4, 2.26.5, 2.26.6, 2.26.7
	4. Defective outdoor fan motor.	4. Repair or replace motor.	2.4; 2.5.1, 2.5.2, 2.5.3, 2.5.2, 2.26; 2.26.2
	5. Defective fan relay on either indoor or outdoor section.	5. Repair or replace relay.	2.51; 2.54
	6. Too long defrost cycle.	6. Replace time clock, defrost relay, or termination thermostat.	2.59; 2.60; 2.61, 2.62
	7. Defective reversing valve.	7. Replace reversing valve.	2.60

Complaint	Possible Cause	Remedy	Reference
	8. Blower belt slipping.	8. Adjust or replace belt.	2.26; 2.26.2
	9. Indoor or outdoor coil dirty.	9. Clean proper coil.	2.26; 2.26.2; 2.27; 2.27.3, 2.27.6
	10. Dirty indoor air filter.	10. Replace or clean filter.	2.27; 2.27.2
	11. Air bypassing indoor or outdoor coil.	11. Prevent air bypassing coil.	2.63
	12. Air volume too low over indoor or outdoor coil.	12. Increase indoor ductwork or remove restriction from coils.	2.52
	13. Auxiliary heat strips ahead of indoor coil.	13. Locate heat strips downstream of indoor coil.	2.64
Compressor cycles on low pressure control.	1. Refrigerant shortage.	1. Repair leak and charge system.	2.27; 2.27.1, 2.27.4, 2.27.5
	2. Low suction pressure.	2. Increase load (see "Suction pressure low").	
	3. Defective expansion valve.	3. Repair or replace expansion valve.	2.28; 2.28.1, 2.28.2, 2.28.3, 2.28.4, 2.28.5, 2.28.6
	4. Dirty indoor or outdoor coil.	4. Clean coil.	2.27; 2.27.3, 2.26; 2.26.2
	5. Slipping blower belt.	5. Replace or adjust blower belt.	2.26; 2.26.2
	6. Dirty air filter.	6. Clean or replace filter.	2.27; 2.27.2
	7. Ductwork restriction.	7. Increase duckwork.	2.52
	8. Liquid drier or suction strainer restricted.	8. Replace drier or strainer.	2.30; 2.30.1, 2.30.2, 2.30.3, 2.30.4, 2.30.5

Condition	Possible Cause	Corrective Action	Reference
	9. Defrost thermostat element loose or making poor contact.	9. Tighten or increase contact.	2.59; 2.60; 2.61
	10. Air temperature too low for evaporation.	10. Relocate unit or provide adequate air temperature.	2.41; 2.41.1, 2.41.2; 2.43
	11. Defrost cycle too long.	11. Replace time clock, defrost relay, or termination thermostat.	2.61
	12. Defective evaporator fan motor.	12. Repair or replace fan motor or relay.	2.26; 2.26.2; 2.46; 2.46.1, 2.46.2, 2.46.3, 2.46.4, 2.46.5, 2.46.6
Outdoor fan runs but compressor will not.	1. Faulty electrical wiring or loose connection.	1. Repair wiring.	2.15
	2. Defective starting capacitor.	2. Replace starting capacitor.	2.18; 2.18.1
	3. Defective starting relay.	3. Replace starting relay.	2.19; 2.19.1, 2.19.2, 2.19.3, 2.19.4, 2.19.5
	4. Defective run capacitor.	4. Replace run capacitor.	2.18; 2.18.2
	5. Shorted or grounded compressor motor.	5. Replace compressor.	2.4; 2.4.1, 2.4.2, 2.4.3, 2.4.4.
	6. Stuck compressor.	6. Replace compressor.	2.6; 2.6.1, 2.6.2
	7. Compressor over-loaded.	7. Determine and remove overload.	2.4; 2.5; 2.6; 2.6.1, 2.6.2
	8. Defective contactor contacts.	8. Replace contactor or contacts.	2.10; 2.10.2
	9. Single-phasing of three-phase power.	9. Locate problem and repair or contact power company.	2.3; 2.15

	10. Low voltage.	10. Locate and correct cause.	2.17
Outdoor fan motor will not start.	1. Faulty electrical wiring or loose connections.	1. Repair wiring or connections.	2.15
	2. Defective outdoor fan motor.	2. Repair or replace motor.	2.5; 2.5.1; 2.5.2, 2.5.3, 2.5.4, 2.26; 2.26.2; 2.46; 2.46.1, 2.46.2, 2.46.3, 2.46.4, 2.46.5, 2.46.6
	3. Defective outdoor fan relay.	3. Replace fan relay.	2.54
	4. Defective defrost control, timer, or relay.	4. Replace control, timer, or relay.	2.61
Outdoor section does not run.	1. No electrical power. power.	1. Inform power company.	2.1; 2.2; 2.3
	3. Faulty electrical wiring or loose terminals.	3. Repair wiring or terminals.	2.15
	4. Compressor over-loaded.	4. Determine overload and correct.	2.5; 2.6; 2.7; 2.7.1, 2.7.2; 2.53
	5. Defective transformer.	5. Replace transformer.	2.47; 2.47.1
	6. Burned contactor coil.	6. Replace contactor coil.	2.5; 2.5.1
	7. Compressor over-load open.	7. Determine cause and correct.	2.6.5, 2.6.6, 2.6.7
	8. High pressure control open.	8. Determine cause and correct.	2.14; 2.14.2
	9. Low pressure control open.	9. Determine cause and correct.	2.14; 2.14.1
	10. Thermostat off.	10. Turn thermostat on and set.	2.13; 2.13.1, 2.13.2

Condition	Possible Cause	Corrective Action	Reference
Indoor blower will not run.	1. Blown fuse.	1. Replace fuse and correct cause.	2.3
	2. Faulty electrical wiring or loose connections.	2. Repair wiring or connections.	2.15
	3. Burned transformer.	3. Replace transformer.	2.47; 2.47.1;
	4. Indoor fan relay defective.	4. Replace fan relay.	2.51
	5. Faulty indoor fan motor.	5. Repair or replace motor.	2.46; 2.46.1, 2.46.2, 2.46.3, 2.46.5
	6. Faulty thermostat.	6. Replace thermostat.	2.13; 2.13.1, 2.31.2
Indoor coil iced over.	1. Dirty filters.	1. Clean or replace filters.	2.27; 2.27.2
	2. Dirty coil.	2. Clean coil.	2.27; 2.27.3
	3. Blower fan belt slipping.	3. Replace or adjust belt.	2.26; 2.26.2
	4. Outdoor check valve sticking closed.	4. Replace check valve.	2.59
	5. Defective indoor expansion valve.	5. Clean or replace expansion valve.	2.28; 2.28.1, 2.28.2, 2.28.3, 2.28.4, 2.28.5
	6. Low indoor air temperature.	6. Increase temperature.	2.26; 2.26.1, 2.26.2
	7. Shortage of refrigerant.	7. Repair leak and recharge system.	2.27; 2.27.1
Noisy compressor.	1. Low oil level in compressor.	1. Determine reason for loss of oil and correct. Replace oil.	2.27; 2.27.1, 2.27.2, 2.27.3, 2.27.4, 2.27.5, 2.27.6; 2.28; 2.28.2; 2.30; 2.30.1, 2.30.2, 2.30.3, 2.30.4, 2.38.1, 2.38.2, 2.38.3; 2.39; 2.39.1

Complaint	Possible Cause	Repair	Reference
	2. Defective suction and/or discharge valves.	2. Replace valves and valve plate or compressor.	2.7; 2.7.3
	3. Loose hold-down bolts.	3. Tighten.	2.33; 2.34
	4. Broken internal springs.	4. Replace compressor.	2.5; 2.8; 2.34
	5. Inoperative check valve.	5. Repair or replace check valve.	2.59
	6. Loose thermal bulb on indoor expansion valve.	6. Tighten thermal bulb.	2.30; 2.30.1
	7. Improper superheat setting on indoor expansion valve.	7. Adjust superheat.	2.30
	8. Stuck open indoor expansion valve.	8. Clean or replace valve.	2.30; 2.30.5
Compressor loses oil.	1. Refrigerant shortage.	1. Repair leak and recharge system.	2.27; 2.27.1
	2. Low suction pressure.	2. Increase load on evaporator.	2.27; 2.27.1, 2.27.2, 2.27.3, 2.27.4, 2.27.5, 2.27.6
	3. Restriction in refrigerant circuit.	3. Remove restriction.	2.30; 2.30.1, 2.30.2, 2.30.3, 2.30.4, 2.30.5
	4. Indoor expansion valve stuck open.	4. Clean or replace expansion valve.	2.28; 2.28.2
Unit operates normally in one cycle, high suction pressure on other cycle.	1. Leaking check valve	1. Replace check valve.	2.59
	2. Loose thermal bulb	2. Tighten thermal bulb.	2.28; 2.28.1
	3. Leaking reversing valve.	3. Replace reversing valve.	2.60
	4. Expansion valve stuck open on indoor or outdoor coil.	4. Repair or replace expansion valve.	2.28; 2.28.2

Condition	Possible Cause	Corrective Action	Reference
Unit pumps down in cool or defrost cycle but operates normally in heat cycle.	1. Defective reversing valve.	1. Replace reversing valve.	2.60
	2. Defective power element on indoor expansion valve.	2. Replace power element.	2.28; 2.28.5
	3. Restriction in refrigerant circuit.	3. Locate and remove restriction.	2.30; 2.30.1, 2.30.2, 2.30.3, 2.30.4, 2.30.5
	4. Cogged indoor expansion valve.	4. Clean or replace expansion valve.	2.28; 2.28.2, 2.28.3, 2.28.4, 2.28.5
	5. Check valve in outdoor section sticking closed.	5. Replace check valve.	2.59
Head pressure high.	1. Overcharge of refrigerant.	1. Purge overcharge.	2.26; 2.26.6
	2. Air or non-condensables in	2. Purge non-condensables.	2.26; 2.26.7
	3. High air temp erature supplied to condenser.	3. Reduce air temperature	2.48
	4. Dirty indoor or outdoor coil.	4. Clean coil.	2.26; 2.26.1; 2.27; 2.27.3
	5. Dirty indoor air filter.	5. Clean or replace filter.	2.27; 2.27.2, 2.27.3
	6. Indoor or outdoor blower belt slipping.	6. Replace or adjust blower belt.	2.26; 2.26.2
	7. Air bypassing indoor or outdoor coil.	7. Prevent air bypassing coil.	2.62
Suction pressure high.	1. Defective compressor suction valves.	1. Replace valves and valve plate or compressor.	2.7; 2.7.3

	2. High head pressure.	2. See previous entry "Head pressure high."	
	3. Excessive load on cooling.	3. Determine cause and correct.	2.53
	4. Leaking reversing valve.	4. Replace reversing valve.	2.60
	5. Leaking check valve.	5. Replace check valve.	2.59
	6. Indoor or outdoor expansion valve stuck open.	6. Clean or replace expansion valve.	2.27; 2.27.2
	7. Loose thermal bulb on indoor or outdoor expansion valve.	7. Tighten bulb.	2.27; 2.27.1
Suction pressure low.	1. Shortage of refrigerant.	1. Repair leak and recharge unit.	2.27; 2.27.1
	2. Blower belt slipping on indoor or outdoor unit.	2. Replace belt or adjust.	2.26; 2.26.2
	3. Dirty air filters.	3. Clean or replace.	2.27; 2.27.2
	4. Defective check valve.	4. Replace check valve.	2.59
	5. Restriction in refrigerant circuit.	5. Locate and remove restriction.	2.29; 2.29.1, 2.29.2, 2.29.3, 2.29.3, 2.29.4, 2.29.5
	6. Ductwork small or restricted.	6. Repair or replace ductwork.	2.52
	7. Defective expansion valve power element on indoor or on indoor or outdoor coil.	7. Replace power element.	2.28; 2.28.1, 2.28.2, 2.28.3, 2.28.4, 2.28.5
	8. Clogged indoor or outdoor expansion valve.	8. Clean or replace valve.	2.28; 2.28.5

Condition	Possible Cause	Corrective Action	Reference
	9. Wrong superheat setting on indoor or outdoor expansion valve.	9. Adjust superheat setting.	2.28; 2.28.1, 2.28.2, 2.28.3,
	10. Dirty indoor or outdoor coil.	10. Clean coil.	2.27; 2.27.3; 2.26; 2.26.2
	11. Bad contactor contacts.	11. Replace contactor or contacts.	2.10; 2.10.2
	12. Low refrigerant charge.	12. Repair leak and recharge system.	2.27; 2.27.1

CHAPTER 2

Component Troubleshooting

This book is intended to provide assistance to the service engineer for diagnosing problems and repairing air conditioning equipment. Through the use of this book, more competent and economical service of equipment should be attained.

The problems encountered during the servicing of air conditioning equipment can be described as follows:

1. When the refrigeration unit does not operate at all, the problem is in the electrical circuit.
2. When the refrigeration system will not refrigerate, the problem is in the mechanical components.
3. The problem can be the result of a combination of electrical and mechanical malfunctions.

If the service technician can determine whether the trouble is electrical or mechanical in nature, he has eliminated half the possible causes of the problem. An example of a combination of electrical and mechanical problems would be if the bearings in a compressor became stuck and caused the compressor motor to burn out.

2.1 POWER FAILURE

When an electrical power failure occurs, the electric company must be contacted to make the necessary repairs. The repairs may consist of replacing a fuse on the electric pole, replacing a transformer, or any one of many other such repairs. However, it is the electric company that must make the repairs to the power supply. After the repairs have been made, the air conditioning service technician should check all of the equipment to be certain that it is operating properly. Many times when an electrical power failure occurs there will be damage to the electric motors, and they will need to be repaired or replaced.

2.2 DISCONNECT SWITCHES

These devices are electrical switches that are used to interrupt electrical power equipment requiring heavy current flow, such as condensing units, large electric motors, and other heavy current using devices. These switches are usually installed within an arm's reach of the equipment that they control. Some of these switches contain only a set of electrical contacts, whereas others contain fuses to the equipment in addition to the contacts. Located on the outside of the switch box is a lever that is used to open and close the contacts (see Figure 2-1).

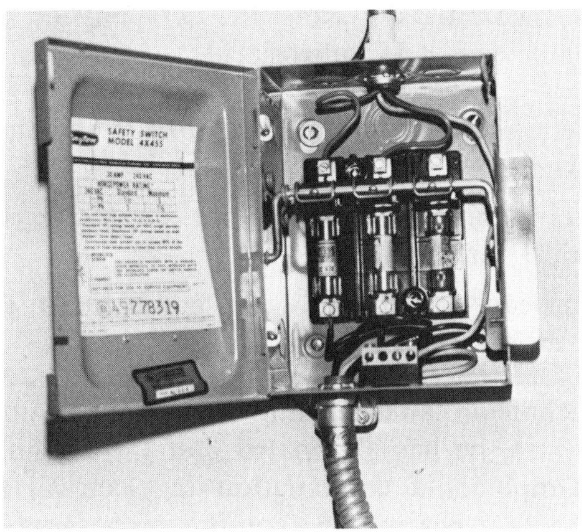

Figure 2–1 Mounted disconnect box.

This lever may be accidentally bumped, causing the contacts to open, or it may accidentally be left in the "off" position. In either case the lever must be returned to the "on" position before equipment operation can be resumed.

2.3 FUSE

An electrical fuse is a protective device placed in the electrical line to an electric circuit. There are basically two types of fuses: the plug type and the cartridge type (see Figure 2-2). The purpose of a fuse is to protect the electric circuit in case of an electrical overload.

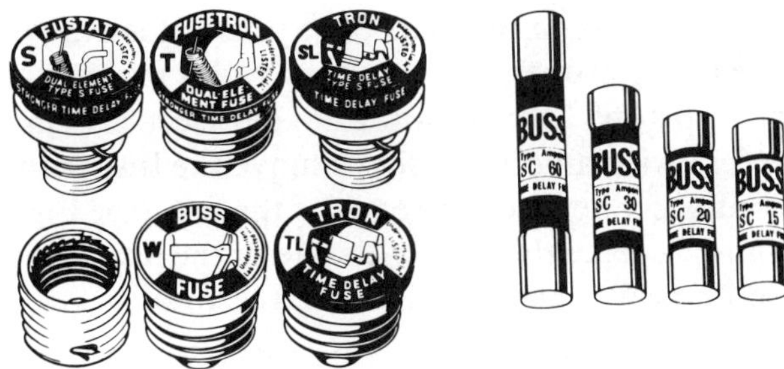

Figure 2–2 Plug and cartridge-type fuses.

This overload may be due to tight bearings, a shorted motor winding, breakdown of electrical insulation, fan belt too tight, burned contactor contacts, etc. The problem that caused the blowm fuse must be found and eliminated before the job is complete.

Defective fuses may be found with either a voltmeter or an ohmmeter. To check fuses with a voltmeter, select a scale that is high enough to prevent damage to the meter. Check the voltage across each fuse (see Figure 2-3).

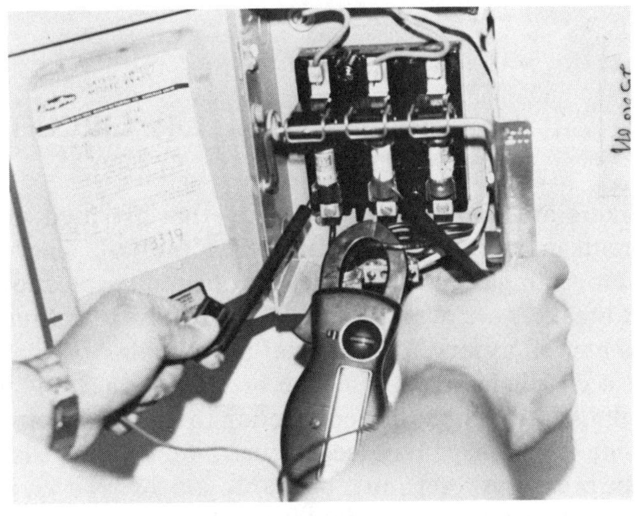

Figure 2–3 Checking fuses in disconnect.

If voltage is found between the ends of any fuse, it is blown and must be replaced with one of the proper size. Obviously, this procedure requires the electrical power to be on. Use caution to avoid electrical shock.

To check the fuse with an ohmmeter, remove the fuse from the holders, and check for continuity between the ends of the fuse (see Figure 2-4). If no continuity is found, replace the fuse with one of the proper size. Be sure to turn off the electric power before removing the fuse.

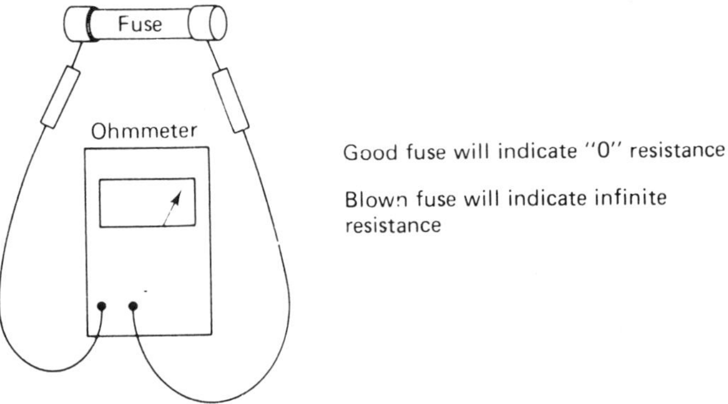

Figure 2–4 Checking fuses out of disconnect box with ohmmeter.

2.4 CIRCUIT BREAKER

Circuit breakers are also protective devices that are used to protect the electrical circuit in much the same manner as the fuse. They may also be used for disconnecting the circuit as well as for circuit protection (see Figure 2-5). Circuit breakers are available in ratings for use with small household refrigerators as well as large commercial installations.

To check a circuit breaker, reset the breaker to the "on" position and check the voltage between the line connection to the breaker and the ground terminal inside the breaker box (see Figure 2-6). If voltage is indicated at this point, the breaker is reset and is probably good. However, if the breaker continues to trip, check the current draw through the circuit with the ammeter. If the current draw is within the rating of the circuit breaker and it continues to trip, replace the breaker; it is weak.

Figure 2–5 Circuit breaker.

Figure 2–6 Checking circuit breaker inside box.

2.5 COMPRESSOR

The compressor is the component used to circulate the refrigerant through the system (see Figure 2-7). The two functions of the compressor are (1) to draw the refrigerant vapor from the evaporator and lower the pressure of

a

b

Figure 2–7 (a) semi-hermetic compressor, (b) hermetic compressor.

the refrigerant in the low side of the system to the desired operating temperature, and (2) to raise the pressure of the refrigerant vapor in the condenser high enough so that the saturation temperature is higher than the temperature of the cooling medium used to cool the condenser and condense the refrigerant.

The compressors in use today are usually of the hermetic or semihermetic type. Therefore, the problems encountered could be electrical, mechanical, or a combination of the two.

2.6 COMPRESSOR ELECTRICAL PROBLEMS

The electrical problems that may be experienced with a compressor can be divided into the following classifications: open winding, shorted winding, or grounded winding. An accurate ohmmeter is needed to check for these conditions. The following checks are good for any type of electric motor.

2.6.1 Open Compressor Motor Windings

These occur when the path for electrical current is interrupted. This electrical interruption occurs when the wire insulation deteriorates and allows the wire to overheat and burn apart. The first step in checking for an open winding is to remove all the external wiring from the compressor terminals. Then use the ohmmeter to check for continuity from one terminal to another (see Figure 2-8).

The ohmmeter should be zeroed before attempting to check motor windings. An open winding will be indicated by an "infinity" resistance reading between any two terminals on the motor. There should be no continuity from any motor terminal to the compressor case.

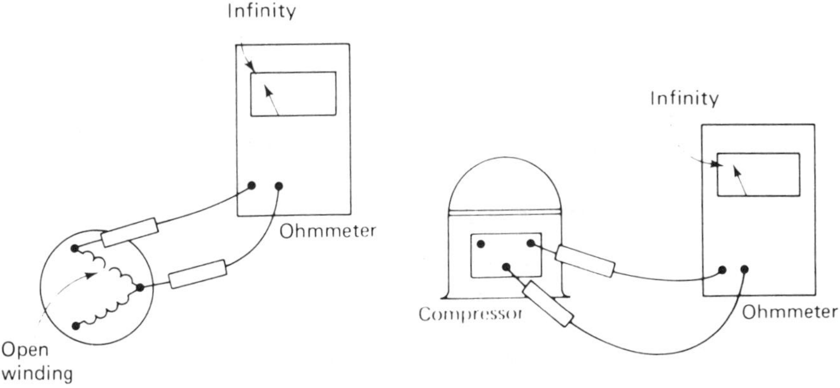

Figure 2–8 Checking for open compressor motor windings.

2.6.2 Shorted Compressor Motor Windings

These occur when the insulation on the winding deteriorates and allows a shorted condition (two wires touch). This condition allows the electrical current to bypass a part of the winding (see Figure 2-9).

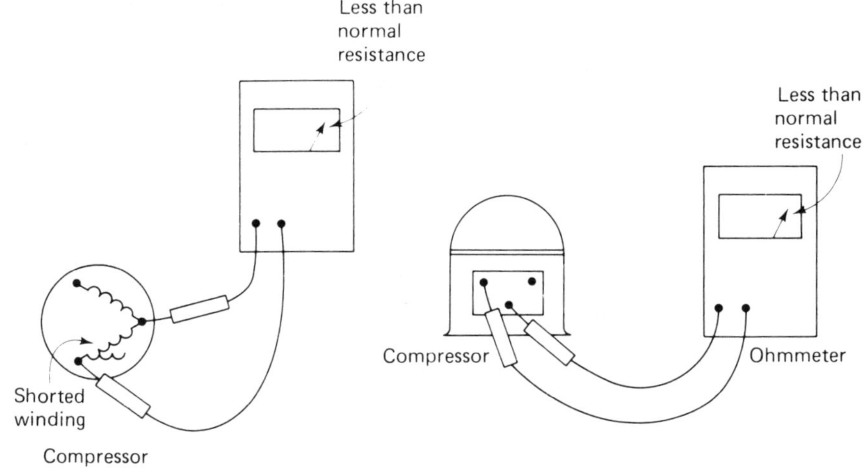

Figure 2–9 Checking for shorted compressor motor windings.

In some instances, depending on how much of the winding has been bypassed, the motor may continue to operate, but it will draw excessive amperage. The first step in checking for a shorted winding is to remove all external wiring from the motor terminals. Then use an ohmmeter to check the continuity from one terminal to another (see Figure 2-10).

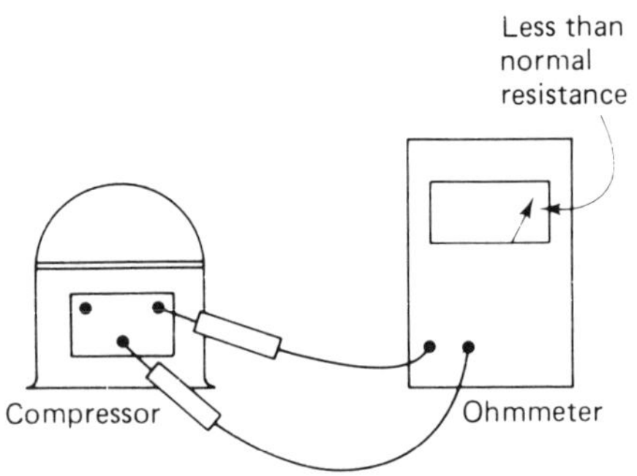

Figure 2–10 Checking compressor motor winding resistance.

The ohmmeter should be zeroed before attempting to make this check. The shorted windings will be indicated by a less than normal resistance. Usually it is better to check the motor manufacturer's data for the particular compressor motor being tested to determine the correct resistance requirements. There should never be any resistance indicated from any terminal to the motor case.

2.6.3 Grounded Compressor Motor Windings

These occur when the insulation on the winding has broken down, and the winding touches the compressor housing or some other metal object (see Figure 2-11). When this condition occurs, the motor will rarely run and will immediately trip the circuit breaker or blow the fuse. To check for a grounded winding, remove all external wiring from the compressor motor terminals. Then use an ohmmeter to check for continuity from each terminal to the motor case. The ohmmeter should be zeroed before attempting to make this check (see Figure 2-12).

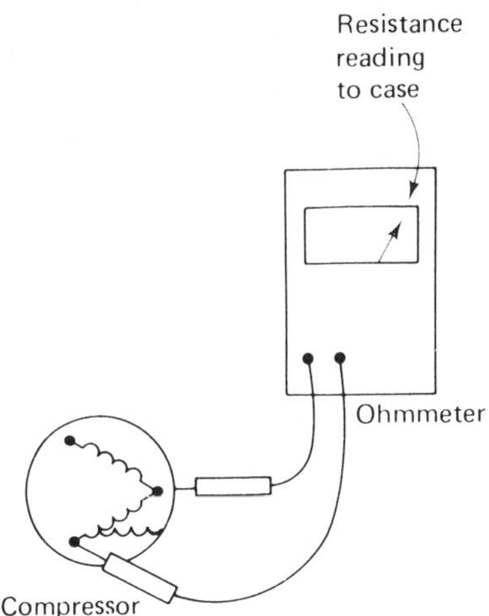

Figure 2–11 Grounded compressor motor winding.

A grounded winding will be indicated by a low resistance reading between one or more terminals and the motor housing. Sometimes, paint or scale must be scraped off the motor housing so that a more accurate reading can be obtained.

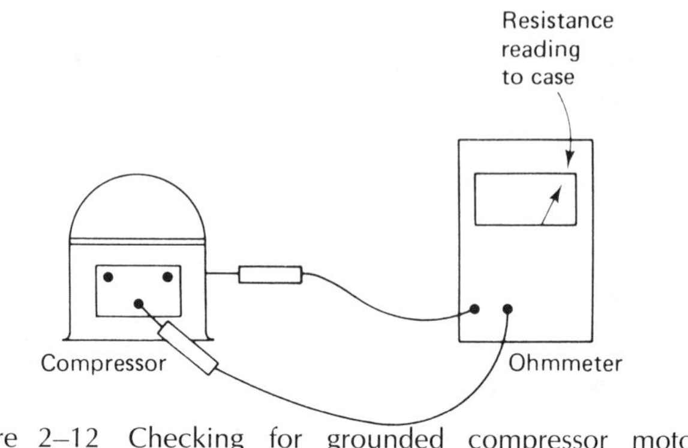

Figure 2–12 Checking for grounded compressor motor winding.

2.6.4 Determining the Common, Run, and Start Terminals

This is a relatively simple process. Make certain that all external wiring is removed from the compressor terminals to prevent false readings on the meter. Draw the terminal configuration on a piece of paper. Then measure the resistance between each terminal with the ohmmeter. Be sure that the ohmmeter has been zeroed before attempting to check continuity. Record the measured resistance on the diagram (see Figure 2-13).

Apply the following formula: the least resistance indicated is between the run and the common terminals; the medium resistance indicated is between the common and the start terminals; the most resistance indicated is between the start and the run terminals. The compressor can now be properly wired.

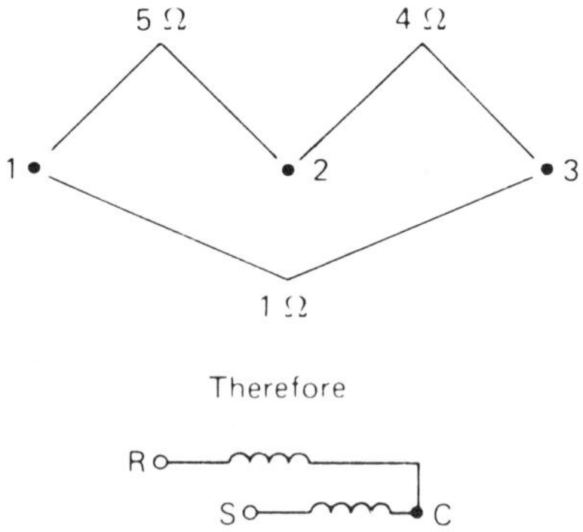

Figure 2–13 Locating compressor terminals.

2.6.5 Checking Two-speed Compressor Windings

This is also a relatively simple process. The first step is to turn off all electrical power to the unit and then remove all external wiring from the compressor motor terminals. Set the ohmmeter on the R x 10K scale. Then check the windings for grounds by touching one probe of the ohmmeter to each terminal and the other probe to the compressor housing. Any resistance reading indicates that there is a grounded winding.

To check for *open windings* in a two-speed single-phase compressor motor refer to Figure 2-14. Check for continuity between terminals 7 and 3, between terminals 7 and 8, and between terminals 7 and 2. When no continuity is indicated with either of these checks, there is an open winding and the compressor must be replaced.

To check for *open windings* in a two-speed three-phase compressor motor refer to Figure 2-15. Check for continuity between terminals 1 and 3 and between terminals 1 and 2. When no continuity is indicated with either of these checks, there is an open winding and the compressor must be replaced.

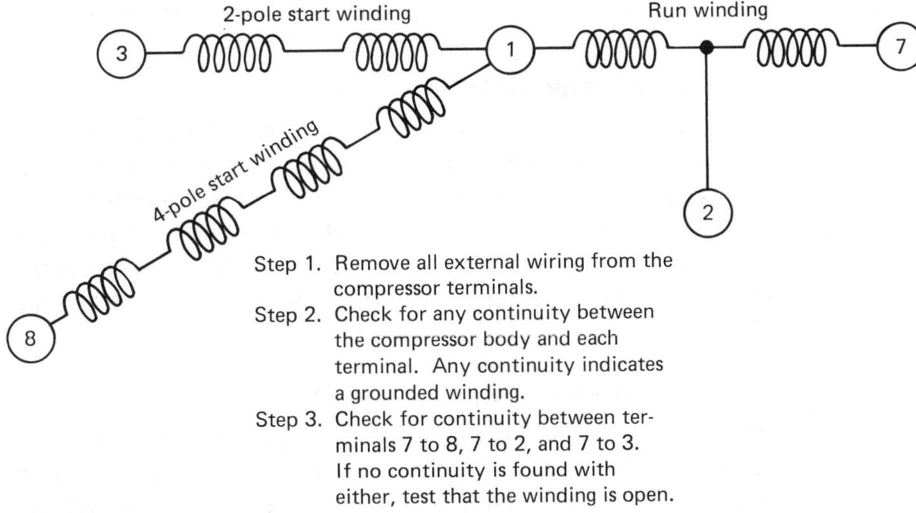

Figure 2–14 Checking motor windings on a two-speed single-phase compressor motor.

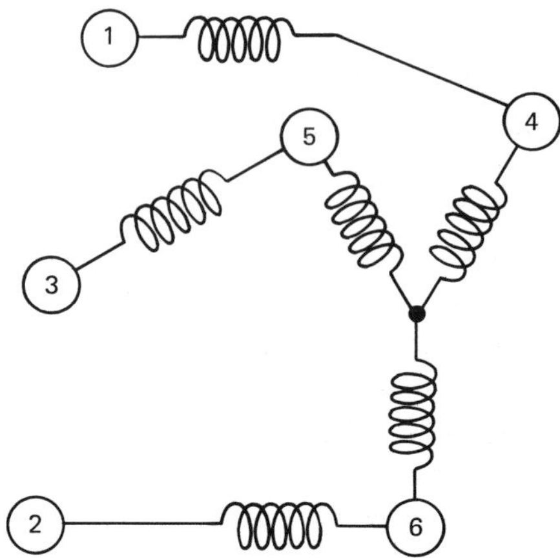

Step 1. Remove all external wiring from the compressor terminals.

Step 2. Check for any continuity between the compressor body and each terminal. Any continuity indicates a grounded winding.

Step 3. Check for continuity between terminals 1 to 2 and 1 to 3. If no continuity is found with either, test that the winding is open.

Figure 2–15 Checking motor windings in a two-speed three-phase compressor motor.

2.6.6 Compressor Motor Overload

These protectors are the devices that are used to protect the compressor motor from damage that may occur due to an overcurrent condition, overtemperature condition, or both. Motor overloads may be mounted inside the compressor housing, or they may be mounted on the outside of the compressor depending on the design of the compressor motor. They are designed to be mounted at the hottest point of the motor winding called the "hot spot."

2.6.7 External Mounted Overload Protectors

These are available in three different types of configurations: (1) the two-terminal, (2) the three-terminal, and (3) the four-terminal (see Figure 2-16). To check the two-terminal overload, place an ammeter on the common electric line to the compressor. Start the compressor while observing the ammeter. The ammeter should indicate a momentary current draw of approximately six times the running amperage of the compressor motor, then drop back to, or below, the rated amperage draw of the motor. If the overload then causes the compressor motor to cycle off, the overload is defective. If the amperage draw remains above the rated amperage of the

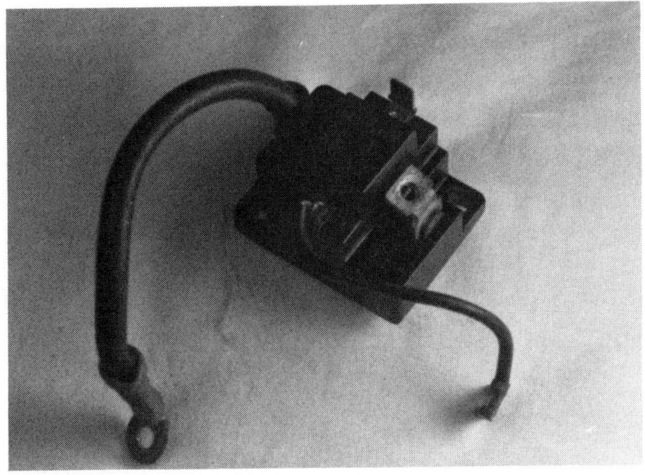

Figure 2–16 Two- , three- , and four-terminal Klixon compressor overload protectors.

motor, the trouble is not in the overload device. To check to make certain that the overload is cycling the compressor motor, check across the overload terminals with a voltmeter while the compressor is off (see Figure 2-17).

If the overload is open, a voltage will be indicated on the meter. A no-voltage reading is an indication that the overload has not opened the electric circuit to the motor. The trouble is with some other component. These overloads should always be replaced with an exact replacement as recommended by the manufacturer to prevent damage to the motor windings.

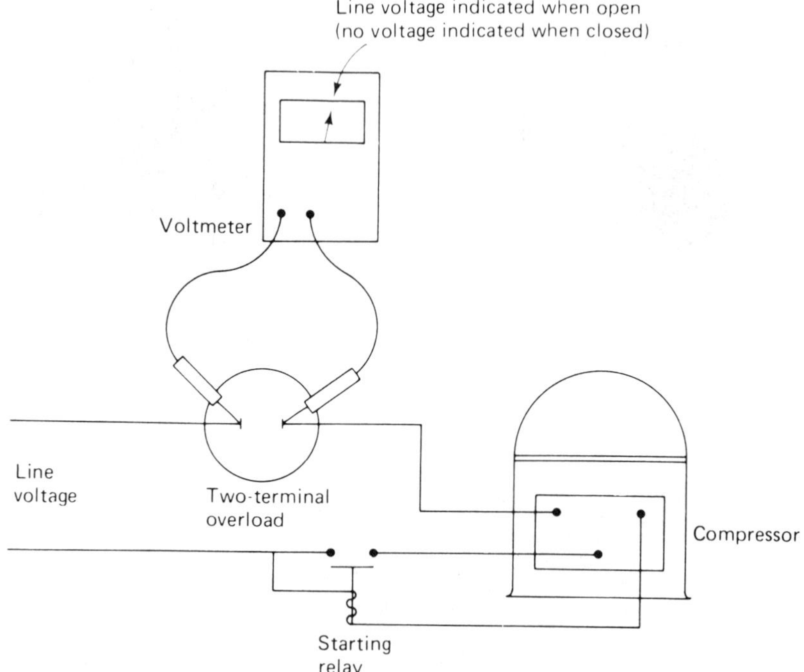

Figure 2–17 Checking voltage across a two-terminal overload.

2.6.8 Three-Terminal External Overloads

These are used on compressor motors where it is desirable to protect the starting winding in addition to the running winding. The terminals are numbered 1, 2, and 3. Terminal 1 is connected to the electrical line going to the compressor common terminal. Terminal 2 is connected to the run terminal of the compressor motor. Terminal 3 is connected to the start capacitor.

Connections of this type will provide closer protection of the compressor motor in the event of a bad starting component. To check a three-terminal overload, place an ammeter on the line connected to terminal 1 and start the compressor while observing the ammeter (see Figure 2-18).

The ammeter should indicate a momentary current flow of approximately six times the normal running current of the motor, then drop back to the amperage rating of the motor or below. If the overload then cycles the compressor motor, the problem is in the overload. If the amperage draw remains above the rated amperage of the motor, the trouble

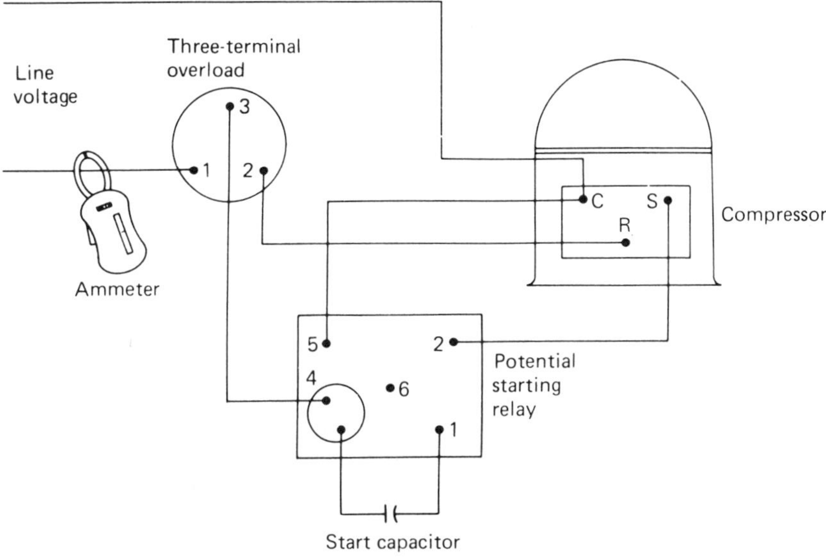

Figure 2–18 Checking amperage draw through a three-terminal overload.

is not in the overload. To check to see if the trouble is in the starting or running circuit, measure the amperage draw through the wire connected to terminal 2, then the wire to terminal 3, while the compressor is running. The circuit with the high amperage draw is where the fault is. The external components must be checked. To be certain that the overload is cycling the compressor motor, check the voltage between terminals 1 and 2, then between terminals 1 and 3, while the compressor is off and cool. If voltage is indicated, replace the overload (see Figure 2-19). If the overload must be replaced, be sure to use an exact replacement to provide the proper protection.

2.6.9 The Four-Terminal Type of External Compressor Motor Overload

Used on larger compressor motors, this is usually mounted away from the compressor in the control panel or on the starter. This type of overload senses only the current draw of the compressor motor (see Figure 2-20). This type of overload is actuated either by a bimetal, by the melting of a special type of solder, or by a hydraulic fluid in a cylinder. The bimetal and solder types will have two electrical connections for the compressor motor and two electrical connections for the control circuit (see Figure 2-21).

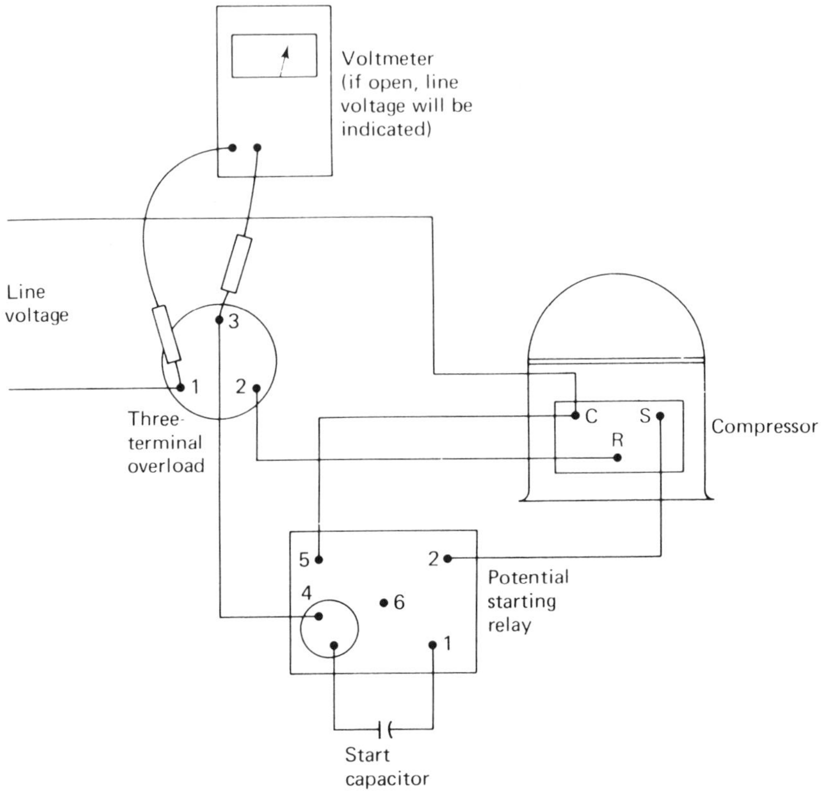

Figure 2–19 Checking voltage across terminals 1 and 3 on a three-terminal overload.

When the current flow to the compressor motor remains over the rating of the overload for a definite period of time, the bimetal or solder pot will become heated and will allow the control circuit to open. This in turn interrupts the control circuit, which in turn stops the compressor motor. To check these overloads, place an ammeter on the compressor motor common wire and start the compressor while observing the ammeter. The ammeter should indicate a momentary flow of approximately six times the running amperage of the compressor motor, then drop back to the rated amperage of the motor or below. If the overload then cycles the motor, the problem is in the overload. If the amperage remains above the rated amperage of the motor, the trouble is not in the overload. To check to make sure that the overload is cycling the compressor motor, check across the control circuit

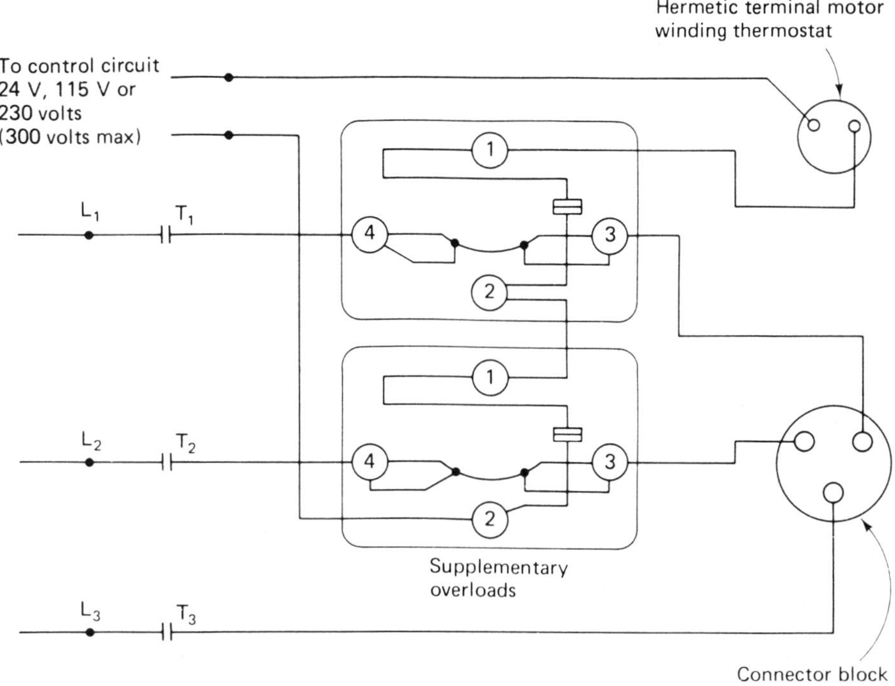

Figure 2–20 External overload connection.

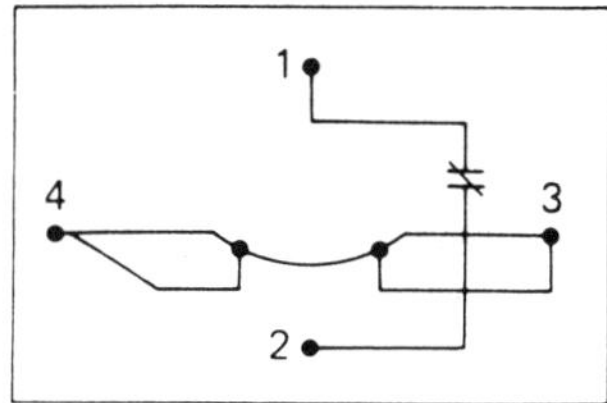

Figure 2–21 Internal view of a four-terminal overload.

terminals of the overload with a voltmeter while the compressor is off (see Figure 2-22).

If the overload is open, a voltage will be indicated. No voltage reading indicates that the overload has not opened the circuit. The trouble is elsewhere. These overloads may be either manual or automatic reset types. Be sure to replace them with an exact replacement so that proper motor protection can be maintained.

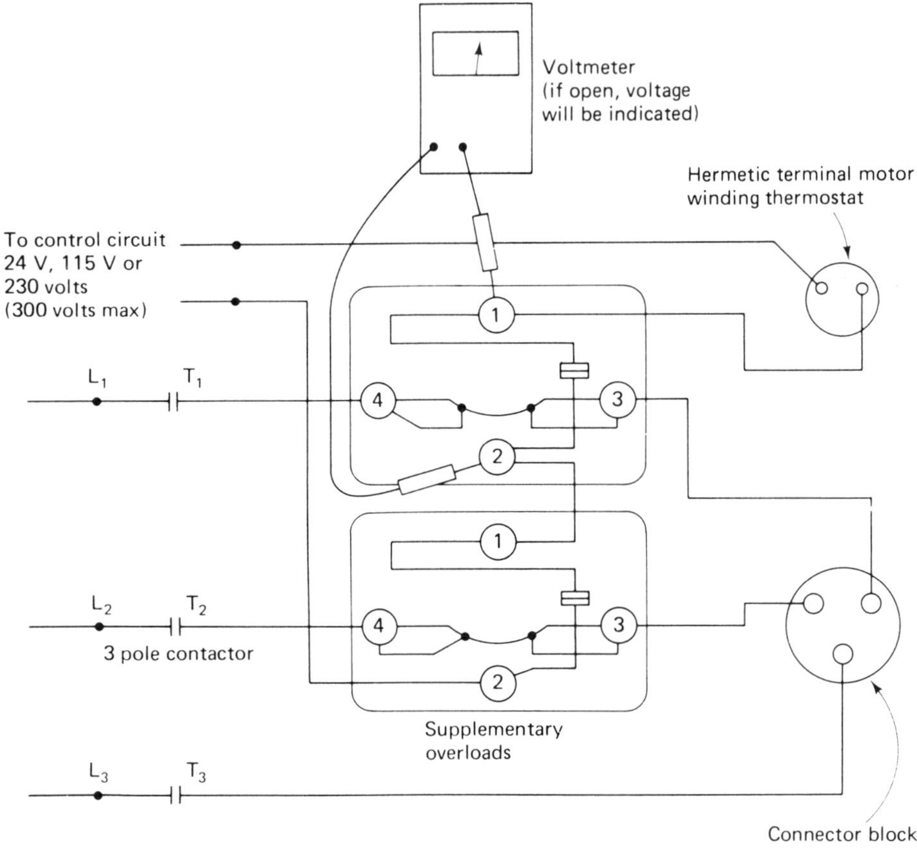

Figure 2–22 Checking voltage across control circuit terminals of a four-terminal overload.

2.6.10 The Hydraulic Fluid-type Compressor Motor Overload

This operates on current flow to the compressor motor only and is generally used on larger compressor motors. It is located in the control panel away from the compressor motor. The current passing through the overload is routed through a heater, which causes the hydraulic fluid to become warm. This warming action increases the pressure of the fluid, causing the electric circuit to the compressor motor to be interrupted. When the hydraulic fluid has cooled down sufficiently, the overload may be reset manually and the compressor started again. To check these overloads, place an ammeter on the wire to the compressor motor common terminal and start the compressor while observing the ammeter (see Figure 2-23).

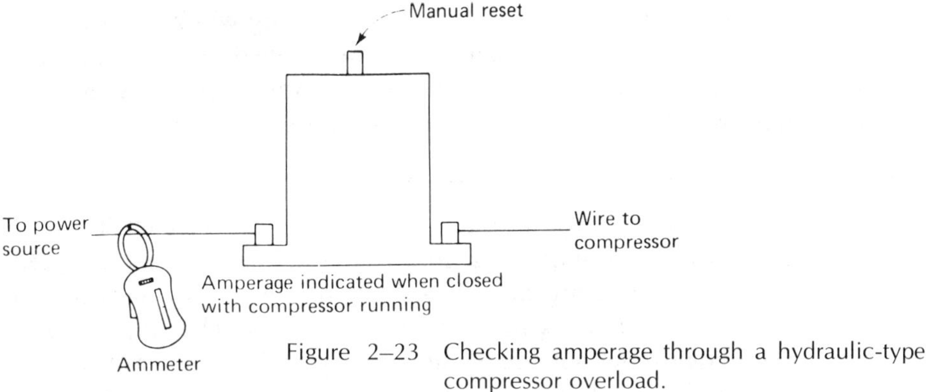

Figure 2–23 Checking amperage through a hydraulic-type compressor overload.

The ammeter should indicate a momentary current flow of approximately six times the rated running amperage of the compressor motor, then drop back to the rated amperage or below. If the overload then cycles the compressor motor, the problem is in the overload. If the amperage draw remains above the rated amperage of the motor, the trouble is not in the overload. To check to make sure that the overload is cycling the compressor motor, check across the line terminals of the overload with a voltmeter while the compressor is off (see Figure 2-24).

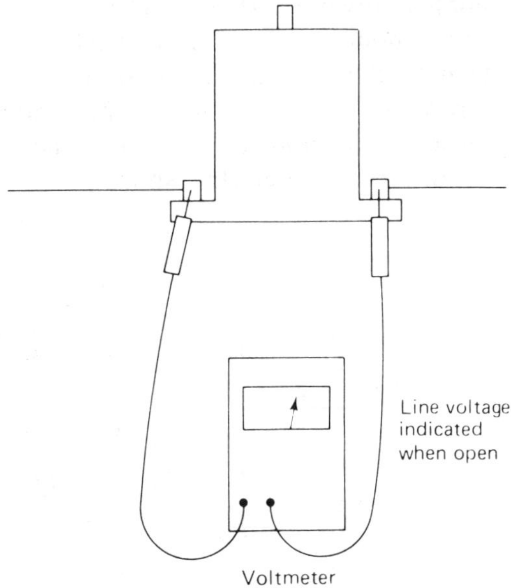

Figure 2–24 Checking voltage across a hydraulic-type compressor overload.

If the overload is open, a voltage reading will be indicated. No voltage indicates that the overload has not opened the circuit. The trouble is elsewhere. These overloads are of the manual reset type. Be sure to replace them with an exact replacement for proper protection.

2.6.11 Internal Motor Thermostats

These function exactly like a two-terminal overload. They are wired in series with the compressor motor contactor coil and are placed inside the compressor housing and imbedded in the motor winding to sense the temperature of the winding more accurately. They have two external line terminals on the outside of the housing and are generally connected to the control circuit. However, some manufacturers place them in the wire to the compressor motor common terminal (see Figure 2-25).

To check these types of thermostats, turn off all electrical power and remove the wiring from the thermostat terminals. Then using an ohmmeter, check across the thermostat terminals with the ohmmeter set on the R × 100 scale. Be sure to zero the ohmmeter. No continuity is an indication that the thermostat contacts are open. This is not, however, an indication that the thermostat is defective. It takes as long as 45 minutes for the thermostat to cool to its normal operating temperature. When it has cooled sufficiently, it will reset automatically. If the compressor housing feels hot to the touch, wait until it has cooled down so that your hand can be held on it comfortably. This is at about 120° F. Then recheck the continuity of the thermostat contacts with the ohmmeter. In some instances, it may be desirable to allow a trickle of water to flow over the compressor housing to aid in the cooling process. *Caution:* do not allow the water to enter the terminal box to prevent a possible electric short.

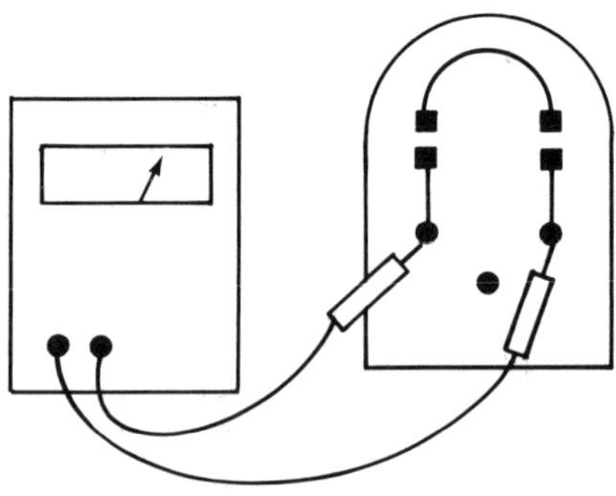

Figure 2–25 Checking internal compressor motor thermostat.

After the compressor has cooled down sufficiently, and only as a last resort, the thermostat terminals can be jumpered. Leave the jumper across the terminals only for a very short period of time to prevent possibly burning out the compressor motor. If the compressor should start, it can be reasonably assumed that the thermostat is defective. In which case the compressor will have to be replaced or the unit operated without this protection. Permanent operation of the compressor with this thermostat jumpered is not recommended, but it can be done until a replacement compressor can be obtained.

2.6.12 Internal Compressor Motor Overloads

These are sometimes used to provide the necessary protection. These controls are different from the internal thermostat because they are line-break devices that have no external wiring connections. The internal overload is embedded into the motor winding much the same as the thermostat, but it is wired in series with the compressor motor common terminal inside the compressor housing (see Figure 2-26). Normally, compressors that are equipped with internal overload protection are marked near the terminal box to indicate their use.

To check the internal overload, remove all electrical leads from the compressor terminals. Then check the continuity with an ohmmeter set on the R × 100 scale between the compressor common and run terminals. Be sure to zero the ohmmeter. If no continuity is indicated, check between the

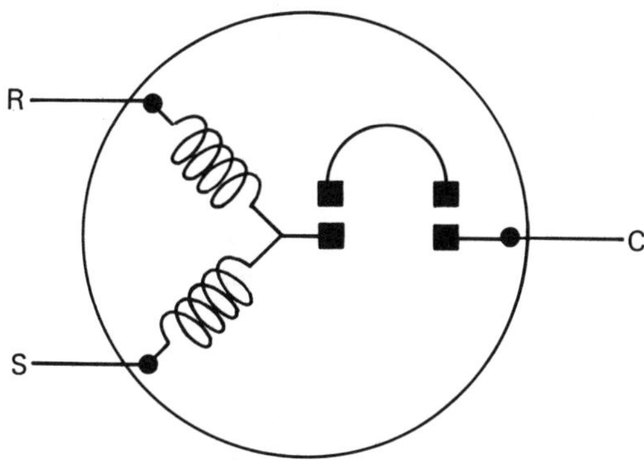

Figure 2–26 Internal overload wiring connections.

start and run terminals for continuity. When no continuity is indicated at either of these points, the overload has tripped. The compressor must be allowed to cool down before the internal overload will reset automatically. A small trickle of water over the compressor housing will often help in cooling the compressor down. To prevent an electrical short, do not allow the water to enter the terminal box. If there is no continuity after the compressor has cooled down to the point where the hand can be comfortably placed on the housing, the overload is defective and the compressor must be replaced before operation of the equipment can be resumed. In any case, make certain that the compressor has had sufficient time to cool before replacing it.

2.7 COMPRESSOR LUBRICATION

Air conditioning compressors, like any moving mechanical device, require lubrication. The proper oil level should be maintained in the crankcase to provide this necessary lubrication. When the compressor is equipped with an oil sight glass, the oil level should be at or slightly above the center of the sight glass (see Figure 2-27).

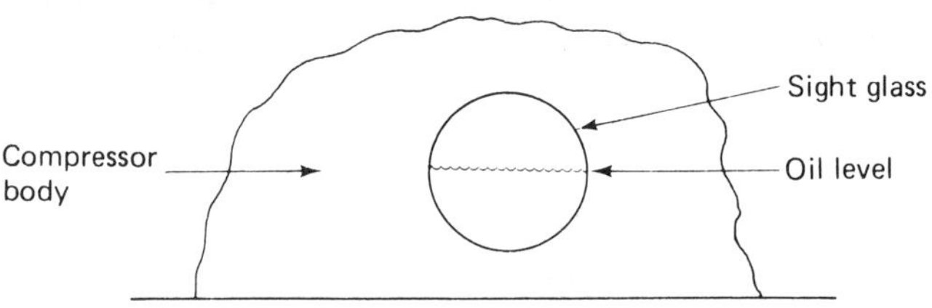

Figure 2–27 Oil level in compresor sight glass.

On compressors not equipped with an oil sight glass, the manufacturer's recommendations should be followed. These recommendations will generally refer to the amount of oil in ounces for a given compressor model. When replenishing the oil in a hermetic compressor, it may be necessary to remove all of the oil from the housing and measure in the correct charge. The proper type of oil for the operating temperature should be used to ensure proper oil return and good compressor lubrication. Refrigeration oil containers must be kept sealed to prevent moisture and other contaminants getting into the oil.

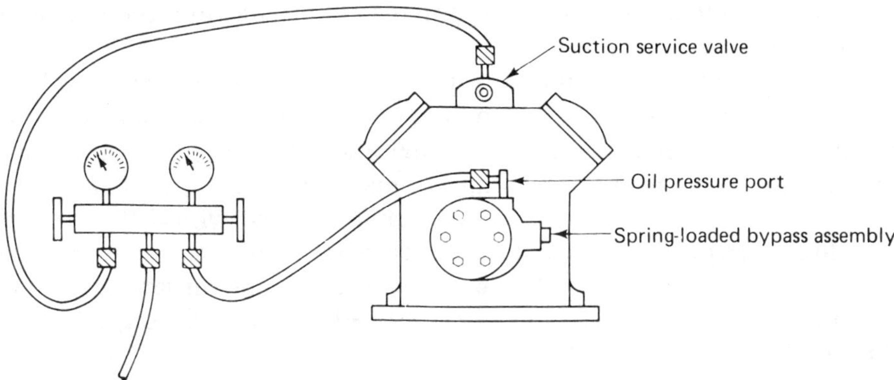

Figure 2–28 Checking oil pressure.

There are three types of lubrication systems used on air conditioning compressors: the splash, pressure, and a combination of both. The splash system is used on compressors up to three horsepower. Compressors of three horsepower and larger are pressure-lubricated. The oil pressure may be checked to determine if proper lubrication is being provided on forced lubrication compressors (see Figure 2-28).
A net oil pressure of 30 to 40 psi (307.09 kPa to 375.79 kPa) is normal; however, adequate lubrication will be provided at pressures down to 10 psi (169.69 kPa). To obtain the net oil pressure, subtract the suction pressure from the oil pump pressure. Example: 90 psig pump pressure –50 psig suction pressure = 40 psi net oil pressure (620.528 Kpa (Kilo-pascal) – 344.738 Kpa = 275.79 Kpa).

2.7.1 Improperly Lubricated Compressor Bearings

These will become tight and sometimes will "freeze" the shaft and prevent operation. If this condition occurs, the compressor motor will become overloaded and will trip the overload or the circuit breaker, or will blow the electrical fuses. In any case, the amperage draw of the compressor motor will be higher than normal. When the compressor motor is not allowed to turn, it will draw locked rotor amperage. This amperage rating is indicated by LR on the motor nameplate. The compressor must be replaced when it is drawing the indicated LR amperage. The compressor oil should be checked, and if it is found to be insufficient, the system must be checked to determine the reason for the lack of oil. The loss of oil may be due to refrigerant leaks, oil logging of the evaporator, low refrigerant charge, etc. The reason must be corrected before the new compressor is placed in operation or it too may soon fail. Do not confuse this condition with faulty starting components or a faulty running capacitor.

A malfunctioning oil pump will generally go undetected until the compressor bearings are damaged enough to cause knocking or until the compressor is frozen mechanically. A malfunctioning oil pump may be due to mechanical wear of the pump. It may become vapor-locked with refrigerant, or the inlet screen may become plugged with dirt or sludge. Should a worn pump be the culprit, it must be replaced at the same time that the compressor is repaired. A vapor-locked oil pump will produce no oil pressure. The vapor lock must be removed by bleeding off the vapor through the gauge connection.

When the oil inlet screen becomes clogged, the oil flow will be restricted and perhaps completely stopped. In this case the screen must either be replaced or cleaned, along with a complete cleaning of the compressor crankcase, replacement of the oil charge, and replacement of the refrigerant filter-drier.

2.7.2 Worn Compressor Bearings

Compressor bearings that are worn become loose and noisy. The compressor loses its efficiency, and the result is poor refrigeration. Loose bearings are usually indicated by more noise than usual from the compressor. In some instances, the bearings will not knock but will cause a vibration in the compressor. When the compressor is equipped with a forced lubrication system, the oil pressure will be lower than normal. The amperage draw will be from normal to low. In some cases, the suction pressure may be high, and the discharge pressure low. When this condition exists, the compressor must be replaced or overhauled depending on the type. This condition is usually the result of age and usage and not faulty lubrication.

2.7.3 Compressor Valves

Compressor valves are the components that control the flow of refrigerant through the compressor. Should they become broken or start leaking, the compressor will become inefficient. A broken or leaking suction valve will result in a higher than normal suction pressure. To check for a defective compressor suction valve, connect the gauge manifold to the service valves on the compressor and open the service valves to the gauge port position (see Figure 2-29).

Next, front-seat the suction service valve (screw it all the way in), and observe the suction pressure while the compressor is operating. The suction pressure should pull down to at least 28-in. vacuum (6.87 kPa) in one or two

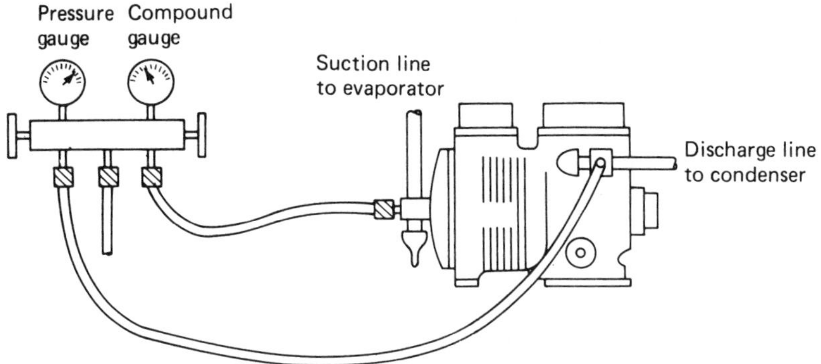

Figure 2–29 Gauges connected to compressor to take refrigerant pressure readings.

minutes. If it does not pump down to this reading, stop the compressor for two or three minutes; then restart the compressor and allow it to run for one or two minutes. If the desired 28-in. vacuum is not reached, the valves must be replaced. When the fault is in open type or semihermetic compressors, the compressor service valuves must be closed, the refrigerant pressure in the compressor relieved, the head removed, and the valve plate replaced (see Figure 2-30).

Figure 2–30 Location of valve plate.

2.7.4 Broken or Leaking Discharge Valves

A broken or leaking discharge valve will result in a lower than normal discharge pressure. To check for a defective compressor discharge valve, connect the gauge manifold to the service valves on the compressor and open the service valves to the gauge port connection (see Figure 2-29). Next, front-seat the suction service valve (screw it all the way in), start the compressor, and allow it to pump as deep as vacuum as possible. Stop the compressor and observe the compound gauge. If the pressure rises, start the compressor and pump another vacuum. Stop the compressor and again observe the compound gauge. If the pressure increases again, close off the discharge service valve. If the pressure on the compound gauge stops rising, the discharge valve is bad and must be replaced. In the case of a hermetic compressor, the compressor must be replaced. When the fault is in open type or semihermetic compressors, the compressor service valves must be closed, the refrigerant pressure in the compressor relieved, the head removed, and the valve plate replaced (see Figure 2-30).

2.8 COMPRESSOR SLUGGING

Slugging is a noisy condition that occurs when the compressor is pumping oil or liquid refrigerant. When a compressor is slugging, it will sound like the clattering of an automobile engine that is under a strain. Continued slugging will probably result in broken valves, scored pistons, and galled bearings. It should be evident that a slugging condition must be corrected.

2.8.1 Oil Slugging of a Compressor

This occurs when there is too much oil in the compressor crankcase, or when the compressor is started with liquid refrigerant in the crankcase. In this situation, some of the oil must be removed to maintain the oil level recommended by the compressor manufacturer. The excess oil may be drained through a drain plug, or some installations may require the removal of the compressor so that the oil may be poured out. In either case the refrigerant must be pumped from the compressor or purged from the system. Do not attempt to remove oil from the compressor with refrigerant pressure in the crankcase.

To pump refrigerant from the compressor, connect the gauge manifold to the compressor service valves (see Figure 2-29). Front-seat the suction service valve and allow the unit to run until approximately 2 psig (114.73 kPa) pressure is shown on the compound gauge. Do not pump the unit below atmospheric pressure for this procedure. Front-seat the compressor discharge service valve and relieve any remaining pressure through the gauge manifold. The compressor may now be serviced.

2.8.2 Refrigerant Slugging

This is the result of liquid refrigerant being returned to the compressor. This can be caused by several things and will usually result in moisture condensing on the compressor housing because of the lower temperature, and sometimes ice or frost will form on the compressor housing. Some of the causes are: overcharge of refrigerant, especially on capillary tube systems; superheat setting too low on the thermostatic expansion valves; automatic expansion valves open too much; or a low-load condition on the evaporator. The obvious solution is to correct any of these causes. However, if these conditions cannot be remedied, a suction line accumulator may be installed to prevent refrigerant slugging of the compressor (see Figure 2-31). Almost all equipment manufacturers also install or recommend the installation of crankcase heaters to reduce slugging due to liquid refrigerant entering the crankcase.

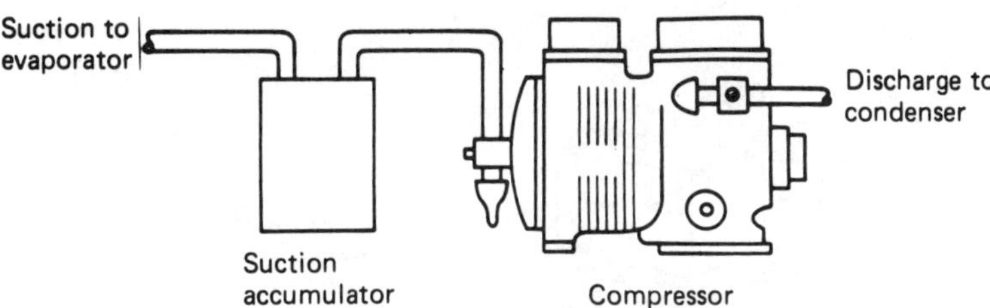

Figure 2–31 Suction line accumulator installation.

2.9 OVERSIZED COMPRESSOR

Generally, an oversized compressor will produce unsatisfactory results. It will produce a lower than normal suction pressure resulting in lower than normal evaporator temperature. This lower than normal temperature causes excessive removal of humidity from the space. This moisture removal is critical and is not desirable in some industrial air conditioning units. The procedure is to replace the compressor with one of the proper size and capacity. However, in some cases it may be possible and less expensive to install a capacity reduction device on the compressor. The manufacturer of the compressor should be consulted to determine whether or not capacity reduction is satisfactory with the particular piece of equipment under question and its use. The manufacturer can also give the proper specifications as to what type is best suited for that unit.

2.10 MOTOR STARTERS AND CONTACTORS

The purpose of a motor starter or contactor is to provide the switching action of a high current and voltage required by the compressor. This is done by a signal given by the control circuit on demand from the thermostat or temperature controller. Motors starters and contractors are electromagnetic-operated devices.

2.10.1 A Burned Starter or Contactor Coil

This will prevent the operation of these devices because there will be no electromagnetic field to operate them. To check for a burned coil, turn off the electrical power to the unit and remove the electric wiring from the terminals of the coil. Check the continuity of the coil with an ohmmeter (see Figure 2-32).

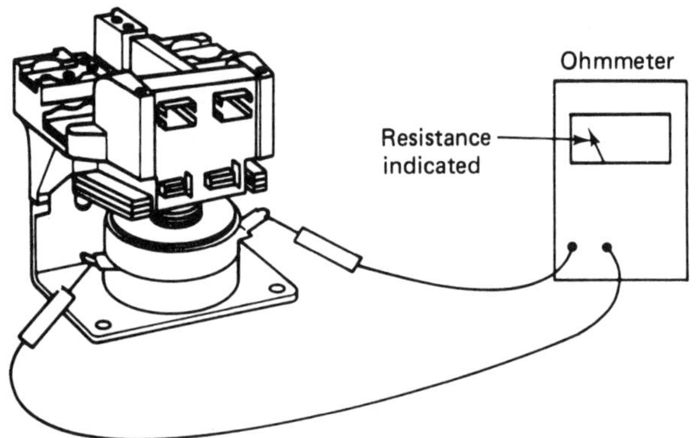

Figure 2–32 Checking continuity of contactor coil.

If the coil is open there should be no continuity indicated on the ohmmeter. Be sure to zero the ohmmeter before attempting to check continuity on any circuit. Many times the insulation on the coil will be discolored, indicating that it has been overheated. When there is no continuity indicated, the coil should be replaced with an exact replacement, or in some cases, the entire starter or contactor must be replaced.

2.10.2 A Sticking Motor Starter or Contactor

This can cause permanent damage to the motor. A starter or contactor that sticks may prevent the motor from starting or it may keep it running when there is no demand for it. When the starter or contactor sticks during the initial start-up, it will usually buzz and either prevent the starting of the motor or cause a delayed starting of the motor. When the starter or contactor sticks closed, the compressor or motor will never stop. There are several types of sprays on the market that may be used to lubricate these troublesome and dangerous controls. However, it is generally recommended that starters and contactors that fall into this category be replaced.

2.10.3 Burned Starter or Contactor Contacts

These can cause permanent damage to the motor windings by preventing the proper flow of current to them. These contacts will be severely pitted and will not make good contact, thus causing a higher current draw than normal (see Figure 2-33). In an emergency, these contacts may be lightly filed until the mating surfaces match (see Figure 2-34). However, the damaged contacts should be replaced as soon as possible because they will again burn and become pitted in a very short time.

Figure 2–33 Burned and pitted contacts.

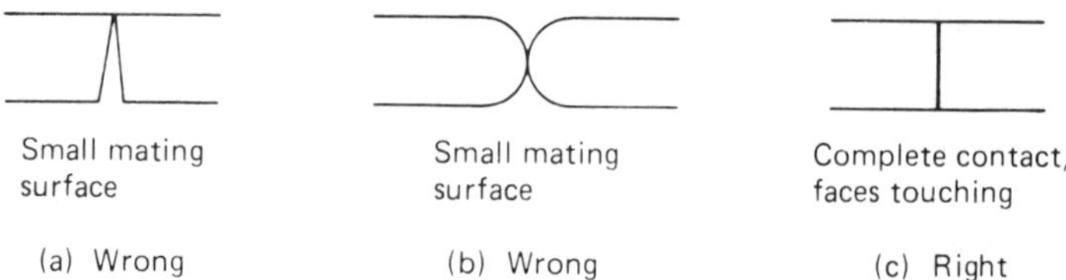

Figure 2–34 Right and wrong contact mating surfaces.

Figure 2–35 Checking voltage across contacts.

Sometimes these contacts may become so bad that they will not make contact at all. This can be determined by energizing the starter or contactor and checking across the contact points with a voltmeter with the electric power on and the contactor closed by demand from the control circuit. Manually closing the contacts may give a false reading (see Figure 2-35). If the contacts are open there will be line voltage indicated on the voltmeter. However, if the contacts are closed, there will be no indication of voltage at this point.

2.11 CONTROL CIRCUIT

The control circuit may be either electrical, electronic, or pneumatic. In the electric and electronic, there is a small current flow through it. The pneumatic control circuit uses air pressure to operate the desired controls. These circuits usually contain various controls, such as the thermostat, contactors, starters, and various safety devices used to protect the compressors, motors, heat exchangers, etc. To check the control circuit, make certain that the electric power to the unit is on and check the transformer or other source of control circuit power; if no power if present there is a blown fuse, a tripped circuit breaker, or a defective transformer that must be replaced. Next, set the thermostat to call for the system to function, then check each component individually for a voltage drop. If the control contacts are closed, no voltage will be indicated on the voltmeter. If the contacts are open, the applied voltage will be indicated (see Figure 3-36). The control where the voltage reading is indicated must be checked further for possible replacement or repair.

2.12 OIL FAILURE CONTROLS

Oil failure controls are used to protect the compressor from a lack of lubrication. The control is actuated by the difference in pressure between the pump outlet and the crankcase pressure. A time delay switch allows the oil pressure to build up to preset operating pressure on compressor start and also prevents nuissance shutdown of the compressor if the oil pressure drops for a short period of time.

To check for a faulty oil failure control, connect a compound gauge to the oil pump outlet. Be sure to leave the oil failure control connected. Connect a voltmeter across terminals 1 and 2 on the oil failure control. Start the compressor and observe the gauges and the voltmeter. The difference in pressure indicated on the two gauges should be at least 10 psi (169.69 kPa) in a short time. When this pressure differential is reached, the contacts between terminals 1 and 2 should open and a voltage indicated on the voltmeter (see Figure 2-37). However, if this minimum pressure is reached and the contacts remain closed, the control will stop the compressor in about two minutes. The control is faulty and must be replaced.

The time delay of these controls is based on 120 or 240 volts as applied in an ambient temperature of 75° F (23.9° C) with the cover in place. If the ambient temperature is much higher than 75° F (23.9° C), the control may be causing nuisance shutdown because of a high ambient temperature rather than low oil pressure. In this case the control must be relocated to a cooler ambient temperature.

2.13 THERMOSTAT

Thermostats are temperature sensitive devices used to control equipment in response to the demands of the space in which they are located. Thermostats

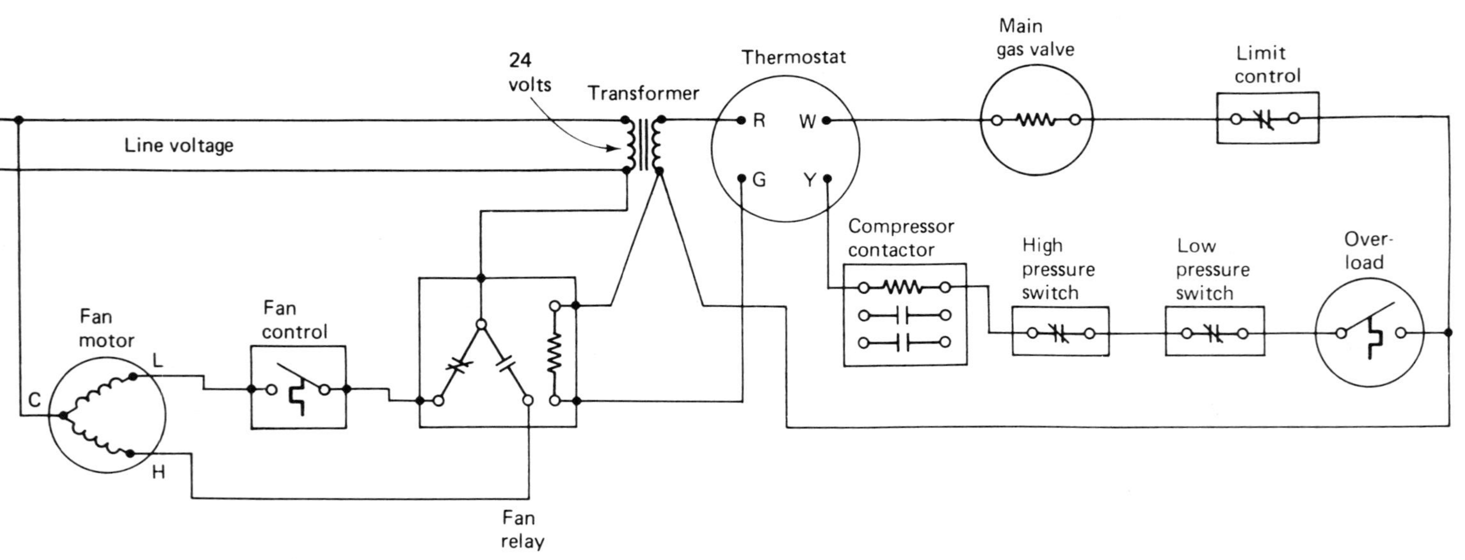

Figure 2–36 Electric heating and cooling control circuit.

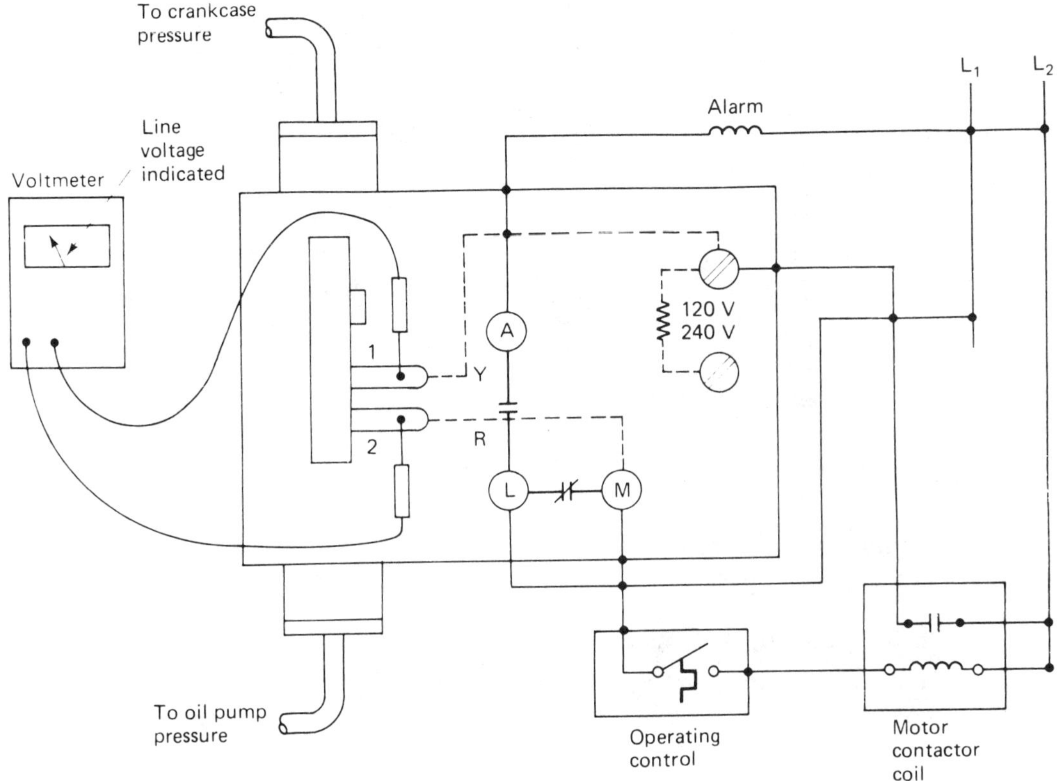

Figure 2–37 Checking operation of oil safety control.

may be operated by a bimetal or "feeler" bulb filled with a fluid that expands and contracts in response to temperature changes.

2.13.1 Room Thermostat

A room thermostat generally makes use of a bimetal element for its operation. This type of thermostat is the most popular for air conditioning and heating applications (see Figure 2-38). To check the thermostat turn it above room temperature and place a reliable thermometer as close as possible to the bimetal element. Allow the thermometer to remain there for ten minutes. Next turn the thermostat temperature selector down. The contacts should "make" (close) at no more than 2°F (1.11°C) below the temperature indicated by the thermometer. If not, the thermostat must be calibrated. There are several means of calibrating a thermostat. Therefore, the manufacturer's specifications should be consulted. If more than 10°F (5.56°C) calibration is needed, replace the thermostat. To check for an in operative thermostat, check the voltage from the red terminal to the cooling

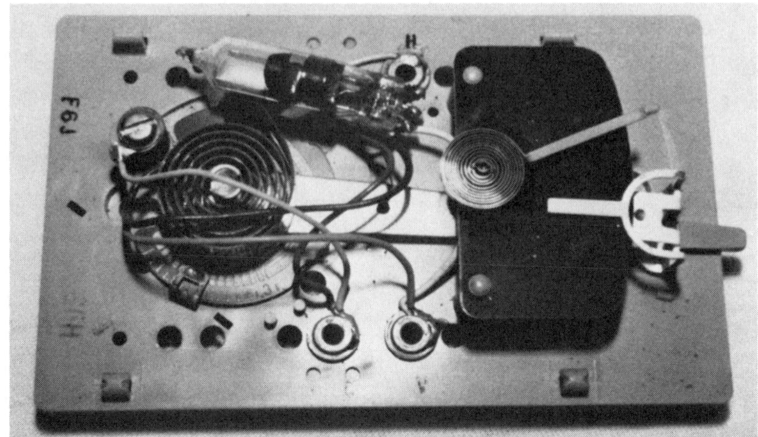

Figure 2–38 Thermostat with cover removed showing bimetal.

terminal. If the contacts are closed, no voltage will be indicated. If the contacts are open, a voltage will be indicated.

2.13.2 Cooling Anticipator

The cooling anticipator incorporated in a thermostat is used during the cooling cycle only. It is not adjustable (see Figure 2-39). Its purpose is to provide a small amount of heat inside the thermostat while the cooling equipment is not in operation. This causes the equipment to start its operating cycle just a little bit sooner than it would without anticipation for moisture removal. When it goes bad the cooling equipment will not operate

Figure 2–39 Cooling anticipator.

properly and will probably result in wide temperature swings inside the space being cooled. When this condition exists, remove the thermostat from the subbase and take an ohm reading between the terminals of the cooling anticipator. If no, or improper, continuity is indicated, replace the thermostat. The manufacturer's specifications will indicate the proper resistance for the model in question.

2.13.3 Thermostat Location

Location of the thermostat is important to satisfactory operation of the equipment. The thermostat should be located on an inside wall about 5 ft. (1.52m) from the floor. It should not be affected by any external heat source such as lights, sun, television, etc. It should be located so that it will sense the average return air temperature.

2.13.4 Thermostat Switches

These are placed according to the type of operation desired. The system switch controls the operation of the equipment. The fan switch controls the operation of the fan. Both switches are usually incorporated in the thermostat subbase (see Figure 2-40). To check out the switches, use a jumper wire to jump from the "R" or "V" terminal to the "Y" or "C" terminal for cooling, or from the "R" or "V" terminal to the "W" or "H" terminal for heating. The first letters of each of these sets refer to Honeywell thermostats; the second refers to General Controls thermostats. When the fan switch is set to "on," the fan will run continuously. The user can select the operation desired. Should these switches become defective, the subbase or thermostat must be replaced before normal operation can be resumed.

Figure 2–40 Air conditioning thermostat.

2.13.5 Outdoor Thermostats

These are remote bulb-type thermostats that are used on heat pump systems. They are sometimes mounted in the terminal box of the outdoor unit. Other times they are mounted under the eave of the building. When mounted under the eave, protection from the wind, rain, and sun must be provided. These thermostats are used to energize the auxiliary heating elements when the outdoor temperature falls below the balance point of the building. The thermostat setting is determined by the designer to provide the greatest efficiency and economy. There may be more than one outdoor thermostat. Therefore, each one is set at a temperature equal to the combined heat output balance point of all the other heat strips plus the heat pump. Electric power for these thermostats is provided through the second stage of the indoor thermostat (see Figure 2-41).

To check the outdoor thermostat, the bulb must be cooled to see if the contacts open and close at the desired temperature. One way to cool the bulb is to insert it in an ice and salt solution along with a thermometer. Adjust the thermostat to the desired temperature if possible. If adjustment is not possible, replace the thermostat.

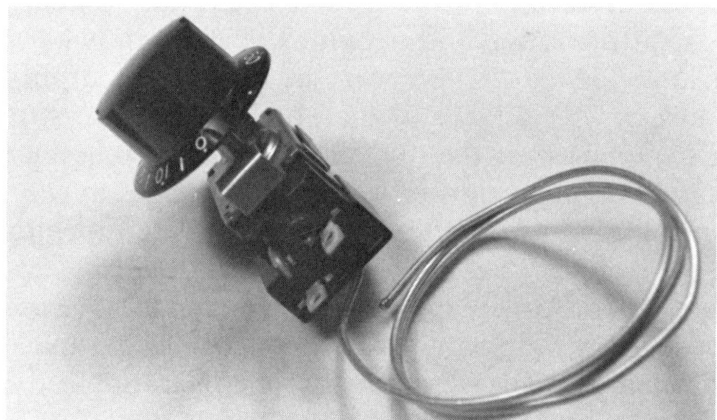

Figure 2–41 Outdoor thermostats.

2.14 PRESSURE CONTROLS

Pressure controls are designed to protect compressors and motors from damage as a result of excessive pressures (see Figure 2-42). Low-pressure controls are used to open the control circuit when the refrigerant pressure in the low side of the system falls below a given pressure. High-pressure controls are used to open the control circuit when the refrigerant pressure in the high side of the system rises to a given pressure. These pressure settings are generally recommended by the equipment manufacturer. Usually when these controls cause the compressor to cycle, the problem is due to some cause other than the control.

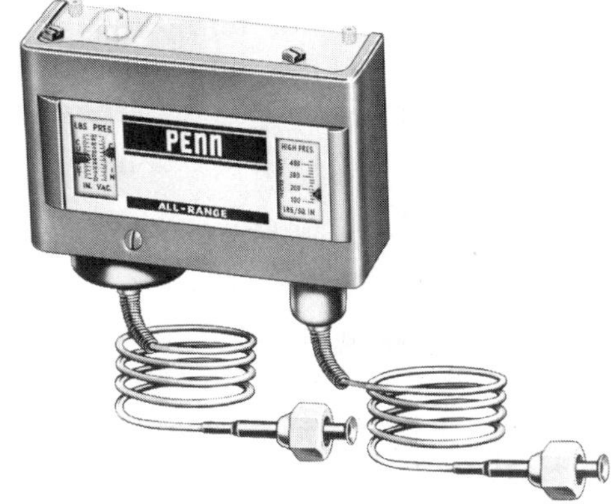

Figure 2–42 Dual Pressure Control.

2.14.1 Low-pressure Controls

Low-pressure controls are designed to respond to the refrigerant pressure in the low side of the system. They will stop the compressor motor to protect it from overheating and to prevent the compressor from pumping oil out of the crankcase. On some smaller units the low-pressure control may also be used as a temperature control. To check the low-pressure control, install a compound gauge on the compressor suction service valve (see Figure 2-43).

Do not disconnect the low-pressure control or cause it to become inoperative. Crack the service valve off the back seat. With the compressor running, front-seat the suction service valve and observe the pressure on the compound gauge when the pressure control stops the compressor. If the actual pressure does not correspond with the desired control setting, adjust

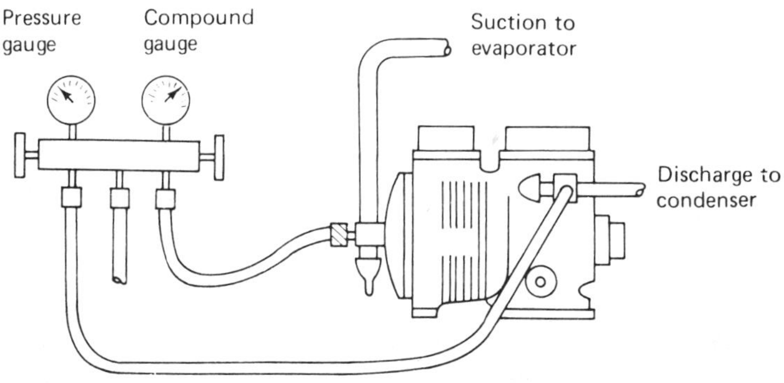

Figure 2–43 Gauges installed on a compressor.

the control, back-seat the suction service valve, and repeat the above procedure until the desired cut-out and cut-in points are obtained. Rarely do these controls need replacement except in the case of refrigerant leakage; in which case, replacement is prefered to repair.

2.14.2 High-pressure Controls

These are designed to respond to the pressure in the high side of the system. They will stop the compressor motor to protect it from being overloaded due to excessively high discharge pressures. To check a high-pressure control, install a pressure gauge on the compressor discharge service valve. Crack the service valve off the back seat. Block the air flow through the condenser, or stop the water pump if it is a water-cooled condenser (see Figure 2–44).

Start the compressor and observe the pressure on the pressure gauge when the control stops the compressor. If the actual pressure does not correspond with the desired control setting, adjust the control, push the reset, and repeat the above procedure until the desired cut-out point is obtained. Rarely do these controls need to be replaced except in the case of refrigerant leakage; in that case, replacement is preferred to repair.

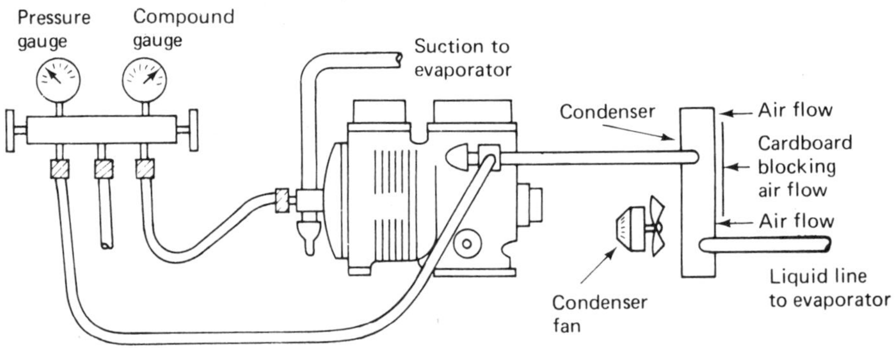

Figure 2–44 Blocking air flow to raise discharge pressure and checking high-pressure control.

2.15 LOOSE ELECTRICAL WIRING

Loose electrical wiring can cause many problems and at times may be extremely difficult to locate. The problems caused by loose electrical wiring do not follow any set pattern. Most loose wiring can be found by visual inspection (see Figure 2-45).

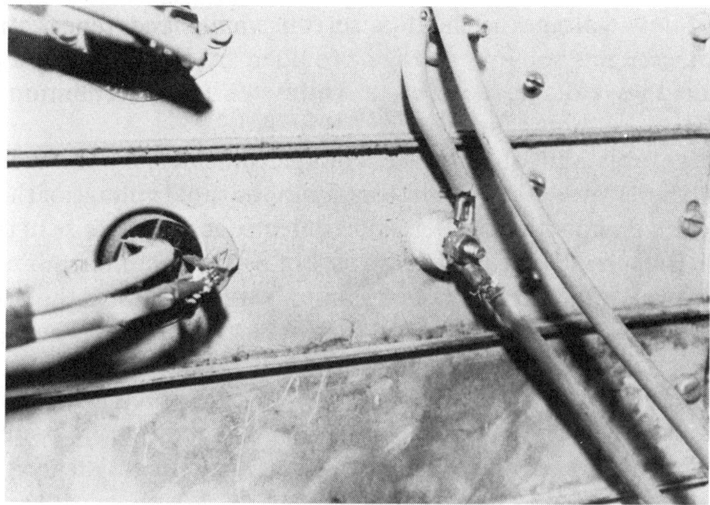

Figure 2–45 Hot electrical joint.

However, when a loose electrical wire is suspected, it is necessary at times to check each wire and its connection individually. This is usually a time-consuming and grueling task, but it must be done before the unit will operate again satisfactorily. Once the bad wire or connection is found, it must be replaced or repaired. Any wiring that is not properly repaired will only cause problems in the future.

2.16 IMPROPERLY WIRED UNITS

Improperly wired units will operate inefficiently, if at all. They will not produce the results desired from the equipment. Each manufacturer designs its own wiring diagrams for each piece of equipment, designs that will cause the unit to produce the desired results. If a service technician is in doubt about the wiring on a piece of equipment, the recommended wiring diagram should be consulted and the wiring changed to match the diagram.

2.17 LOW VOLTAGE

Low voltage to a motor can cause it to overheat and will damage the motor windings. This overheating is due to excessive current draw because of the low voltage. This does not include those types of motors that are designed to operate on a varied voltage, such as some fan motors. There are several causes of low voltage, such as wire too small, loose connections, bad contactor contacts, or low voltage provided by the power company. To check for low voltage, connect a voltmeter to the common and run terminals of the motor (see Figure 2-46).

Start the unit and observe the voltage indicated on the voltmeter. It should not vary more than 10 percent from the rated voltage of the unit. If a voltage drop of more than 10 percent is detected, check the size of the wire to the unit. Be sure that it is at least as big as the recommendation of the equipment manufacturer. If not, it should be replaced with the proper size. If the wire is of sufficient size and the voltage is still low, check for loose connections. These connections will usually be indicated by the wire insulation being overheated or burned. Repair these connections. If there are no loose connections, check the voltage at the electric meter. If it is found to be low here, contact the power company for assistance.

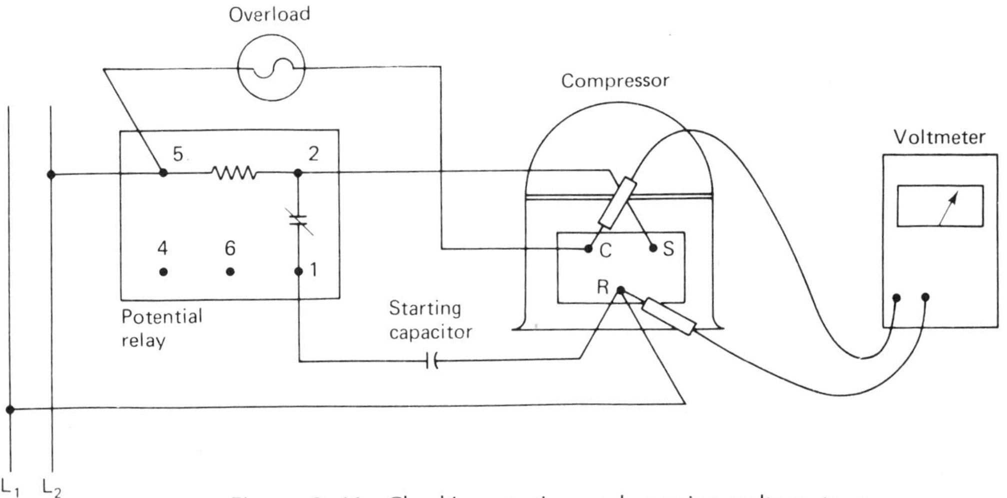

Figure 2–46 Checking starting and running voltage to a motor.

2.18 STARTING AND RUNNING CAPACITORS

Capacitors are used by many manufacturers to improve the starting and running characteristics of their motors. Capacitors are manufactured for use in both starting and running motors. Each manufacturer determines the proper size for its motor; those recommendations should be followed.

2.18.1 Starting Capacitors

These are used in the starting circuit of the motor. They are generally round, encased in a plastic casing, and have a relatively high microfarad (mfd) rating (see Figure 2-47).

Starting capacitors are designed for short periods of use only. Any prolonged use will generally result in damage to the capacitor. If a starting capacitor is found to be defective, be sure to check the starting relay before the unit is placed back into service or the new capacitor may also be damaged. The best way to check a capacitor is to use a capacitor analyzer. This type of meter provides a direct reading of the mfd output of the capacitor without the use of bulky equipment and mathematical formulas (see Figure 2-48).

Starting capacitors that are found to be out of the range of 0 percent to +20 percent of the mfd rating of the capacitor must be replaced with the proper size.

Replacement capacitors must have at least the same or a greater voltage rating as those being replaced. When replacing starting capacitors, be sure that a 15,000-18,000 ohm, two-watt resistor is soldered across the terminals of the capacitor (see Figure 2-47). This is a precautionary measure to prevent arcing and burning of the starting relay contacts. To wire a starting capacitor into the circuit, see Figure 2-49.

Figure 2–47 Starting capacitor with bleed-resistor.

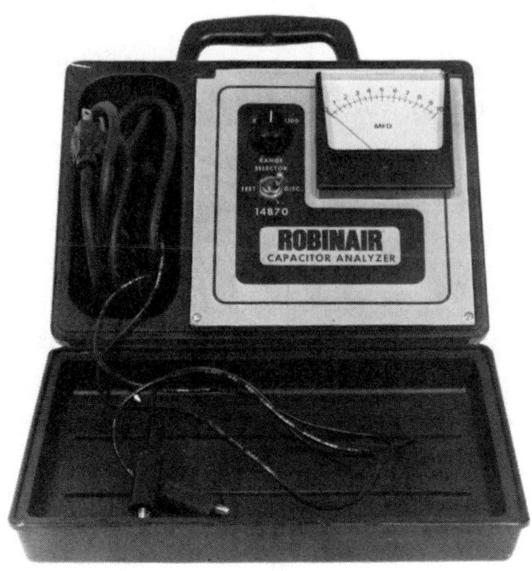

Figure 2–48 Capacitor analyzer (Courtesy of Robinair Manufacturing Co.)

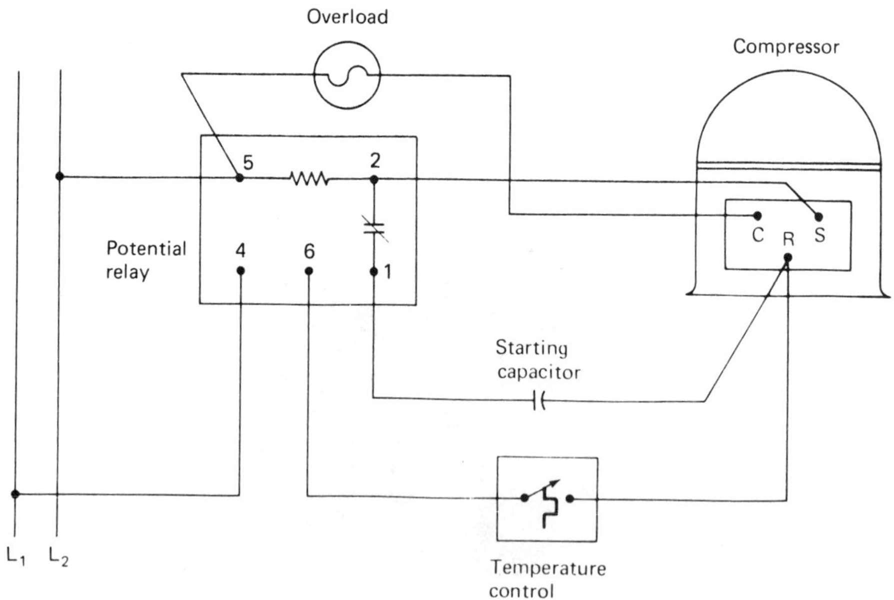

Figure 2–49 Compressor wiring diagram showing starting circuit.

2.18.2 Running Capacitors

Running capacitors are in the operating circuit continuously. They are normally oil-filled type capacitors. The mfd rating of these capcitors is relatively low, even though they are physically large in size. Running capacitors are used to improve the running efficiency of motors. They also provide enough torque to start the PSC (permanent split capacitor) type motors.

Running capacitors are provided with a red dot (see Figure 2-50). The red dot indicates the terminal that is most likely to short out in case of a capacitor breakdown. Because of the relatively high voltage generated in the start winding, the unmarked terminal is connected to the starting terminal on the motor. If the terminal with the red dot is connected to the motor starting terminal, damage to the winding could occur. The best way to check these capacitors is with a capacitor analyzer. This type of meter provides a direct reading of the mfd output of the capacitor without the use of bulky equipment and mathematical formulas (see Figure 2-48). Running capacitors that are found to be out of the range of + or – 10 percent of the mfd rating of the capacitor must be replaced with the proper size. The voltage rating of any capacitor must be equal to or greater than the one being replaced. A higher than normal running amperage usually indicates a weak capacitor. To wire a running capacitor into the circuit, see Figure 2-51.

Figure 2–50 Run Capacitor.

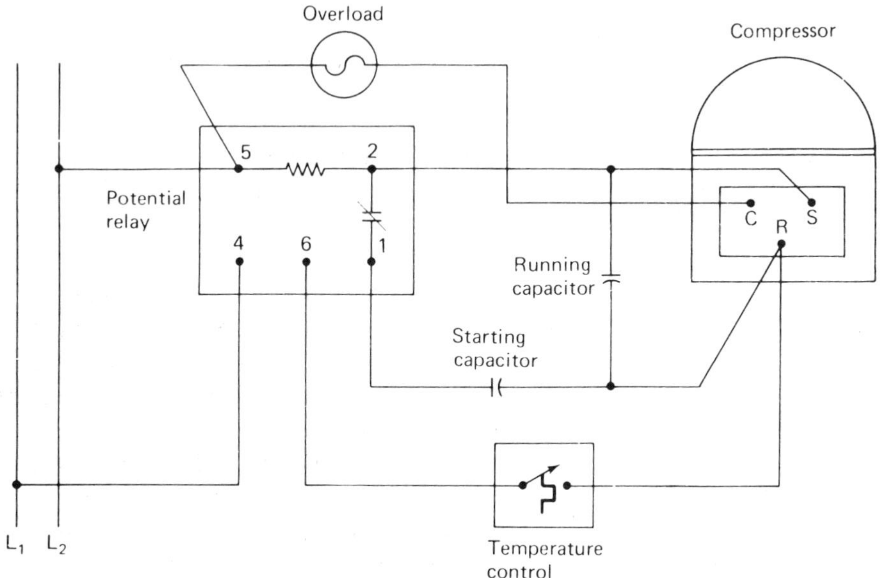

Figure 2–51 Compressor wiring diagram using start and run capacitors.

2.19 STARTING RELAYS

Starting relays are the devices used to remove the starting circuit from operation when the motor reaches approximately 75 percent of its normal running speed. Its function is basically the same as the centrifugal switch used in split-phase motors. There are four types of starting relays in use today: (1) amperage (current) relay, (2) hot wire relay, (3) solid-state relay, and (4) potential (voltage) relay. There is also a solid-state starting device that is used on PSC motor-compressors. The horsepower size and the design of the equipment regulate which type of starting relay is to be used.

2.19.1 The Amperage (Current) Relay

This is an electromagnetic-type relay, which is normally used in ½ hp units and smaller (see Figure 2-52).

These relays are positional types and must be properly mounted for satisfactory operation. They must be sized for each motor horsepower and amperage rating. To check an amperage relay, turn off the electricity to the unit and remove the wire from the "S" terminal and touch it to the "L" terminal on the relay (see Figure 2-53).

Place an ammeter on the common wire to the compressor. Start the compressor and immediately remove the "S" wire from the "L" terminal. If

Figure 2–52 Amperage relay.

the compressor continues to run and the amperage draw is within the amperage rating of the compressor, replace the relay. *Caution:* Do not allow the loose wire to come into contact with anything or anyone to prevent electrical shock.

An amperage relay that is too large for a motor may not allow the relay contacts to close, thus leaving out the starting circuit. The motor will not start under these conditions. A relay that is rated too small for a motor may keep the contacts closed at all times while electrical power is applied to the unit, leaving the starting circuit energized continuously. Damage to the starting circuit may occur under these conditions. A motor protector must be used with this type of relay.

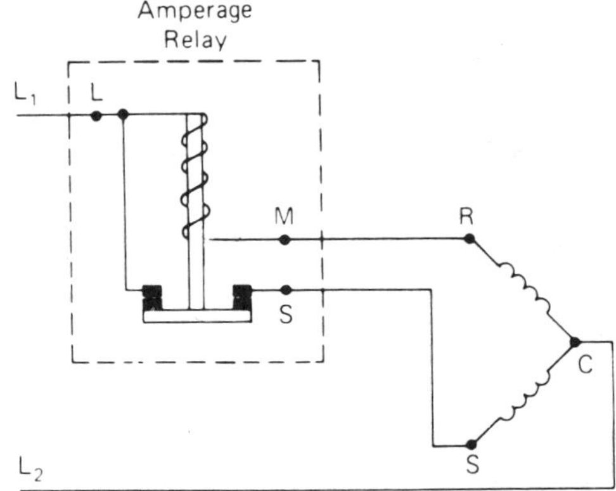

Figure 2–53 Amperage relay connections.

2.19.2 Hot Wire Relays

Hot wire relays are a form of current relay, but they do not operate with an electromagnetic coil. They are designed to sense the heat produced by the flow of electrical current through a resistance wire (see Figure 2-54).

There are two sets of contacts in these relays, a set for starting and a set for running. To check these relays, turn off the electrical power to the unit and remove the wire from the "S" terminal on the relay and touch it to the "L" terminal on the relay (see Figure 2-55). Place an ammeter on the common wire to the compressor. Start the compressor and immediately remove the "S" wire from the "L" terminal. In order to prevent electrical shock, do not allow the loose end of the wire to touch anything or anyone. If the compressor continues to run and the amperage draw is within the amperage rating of the compressor, replace the relay. If the compressor operates within the amperage rating indicated by the manufacturer but still stops within one or two minutes, the overload portion of the relay is defective. Replace the relay with one of the proper size for the horsepower and amperage rating of the compressor motor.

A hot wire relay that is too large for the motor will not remove the starting components from the circuit resulting in possible motor damage. One that is rated too small will stop the motor with the overload after one or two minutes of operation. An additional overload is not necessary when these relays are used. These are nonpositional relays.

2.19.3 Solid-state Starting Relays

Solid-state starting relays use a self-regulating conductive ceramic developed by Texas Instruments, Inc., which increases in electrical resistance as the compressor starts, thus quickly reducing the starting winding current flow to a miliamp level. The relay switches in less than 0.5

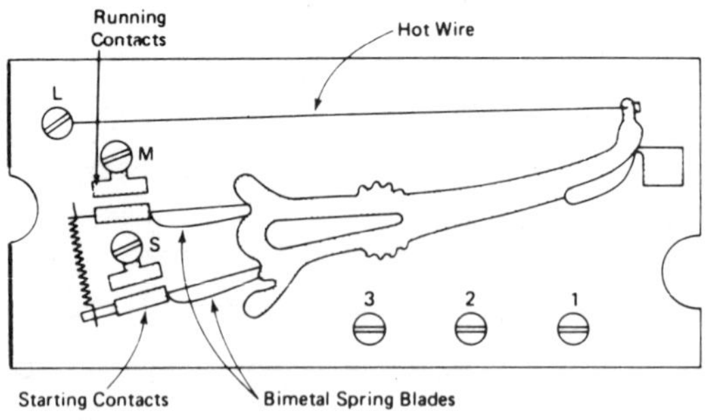

Figure 2–54 Hot wire relay.

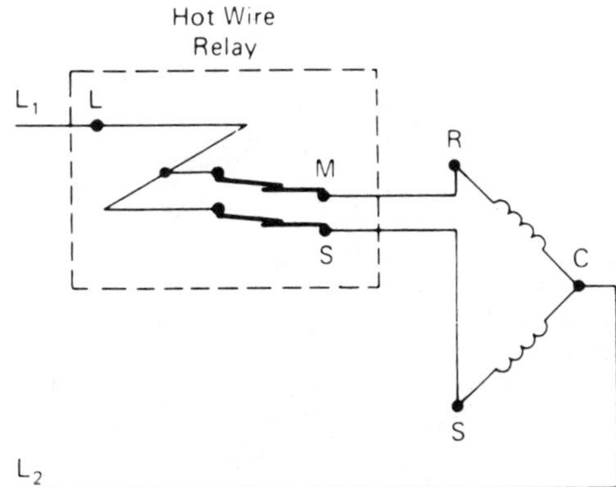

Figure 2–55 Hot wire relay connections.

seconds. This allows this type of relay to be applied to refrigerator compressors without being tailored to each particular system within the specialized current limitations. These relays will start virtually all split-phase 115-volt hermetic compressors up to 1/3 hp. An overload must be used with these relays (see Figure 2-56).

Since these relays are push-on type devices, the easiest method of checking their operation is to simply install a new one. Be sure to check the amperage draw of the compressor motor.

2.19.4 Potential Relays

Potential relays operate on the electromagnetic principle. They incorporate a coil of very fine wire wound around a core. These starting relays are used

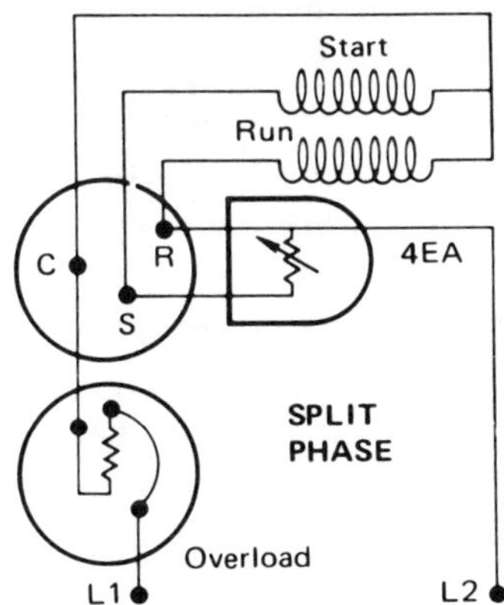

Figure 2–56 Klixon solid-state relay connections. (Courtesy of Texas Instruments).

Figure 2–57 Potential starting relay.

on motors of almost any size. They are nonpositional. The contacts are normally closed and are caused to open when a plunger on the relay is pulled into the relay coil. These relays have three connections to the inside in order for the relay to perform its function. These terminals are numbered 1, 2, and 5. Other terminals numbered 4 and 6 are sometimes used as auxiliary terminals (see Figure 2-57).

To check a potential relay, turn off the electrical current to the unit and remove the wire from terminal 2 on the relay and touch it to terminal 1 on the relay. Place an ammeter on the wire to the common terminal on the motor. Start the motor and immediately remove the #1 terminal on the relay. *Caution:* Do not allow the loose wire to touch anything or anyone to prevent electrical shock. If the compressor continues to run and the amperage draw is within the rating of the compressor, replace the relay.

The sizing of potential relays is not as critical as with the amperage and hot wire relays. A good way to determine what relay is required is to start the motor manually and check the voltage between the start and common terminals on the compressor while the motor is operating at full speed (see Figure 2-58). Multiply the voltage reading by 0.75 and this will be the pick-up voltage of the required relay.

2.19.5 Starting Relay Mounted so that It Vibrates

A starting relay that is mounted so that it vibrates will cause the contacts to arc excessively and become burned. When a relay is mounted on such a

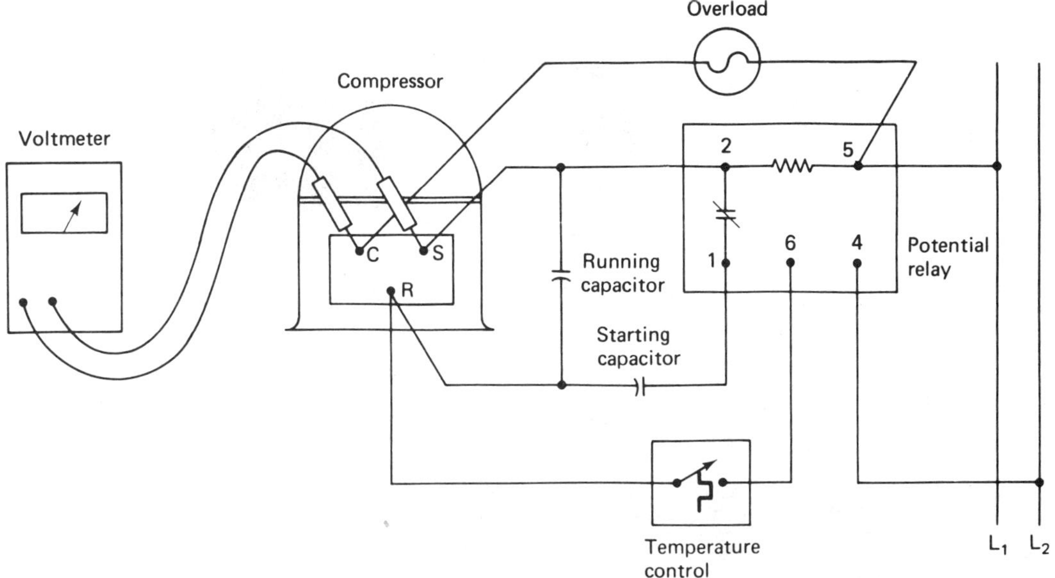

Figure 2–58 Checking voltage between start and common terminals.

surface, it must be remounted on a more solid surface. Generally the relay will need to be replaced during the process. Be sure to replace the relay with the proper size and type, and if it is a positional relay, it must be mounted to satisfy the manufacturer's recommendations.

2.20 POSITIVE TEMPERATURE COEFFICIENT STARTING DEVICE

These PTC resistor devices are used on PSC motors to provide additional starting torque. Their use is not recommended on systems that use a thermostatic expansion valve as the flow control device or on systems that are subjected to short cycling conditions. When they can be used, however, they are very simple to install, less expensive to buy, and have a wide range of application (see Figure 2-59).

The PTC material has a steep slope positive temperature coefficient, which has a cold resistance rating of about 50 ohms and a hot resistance rating of about 80,000 ohms.

The PTC is wired in parallel with the run capacitor and increases the starting torque to about 200 to 300 percent. The PTC material heats up and takes the start-assist out of the circuit in appoximately 1/5 of a second, and the compressor then runs in its normal operating mode.

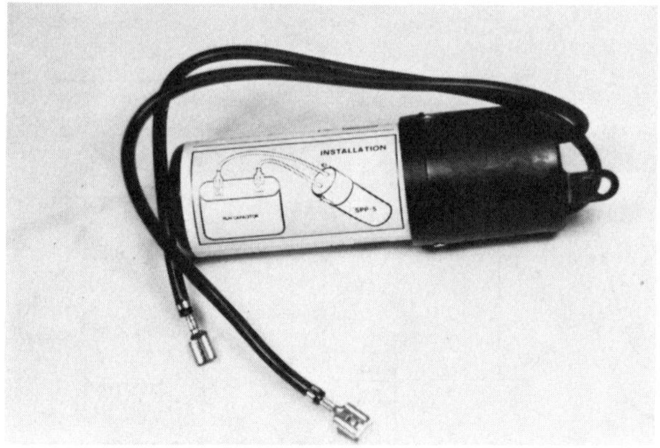

Figure 2–59 Positive temperature coefficient starting device.

To check this start-assist device, allow it to cool to room temperature and check the resistance with an ohmmeter. If it is very far from the cold resistance rating, replace the device.

2.21 CRANKCASE HEATERS

Crankcase heaters are electrical resistors that are designed to provide just enough heat to keep any liquid refrigerant that might enter the crankcase boiled off. There are two different designs: (1) externally mounted, and (2) internally mounted (see Figure 2-60).

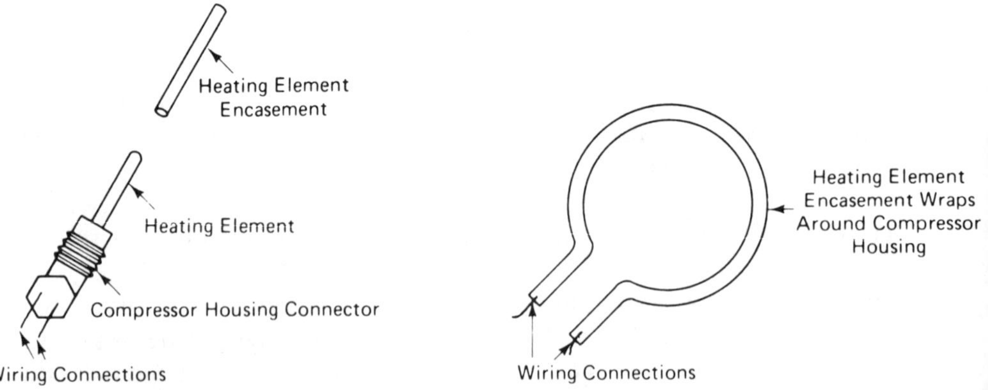

Figure 2–60 (a) Internal crankcase heater, and (b) external crankcase heater.

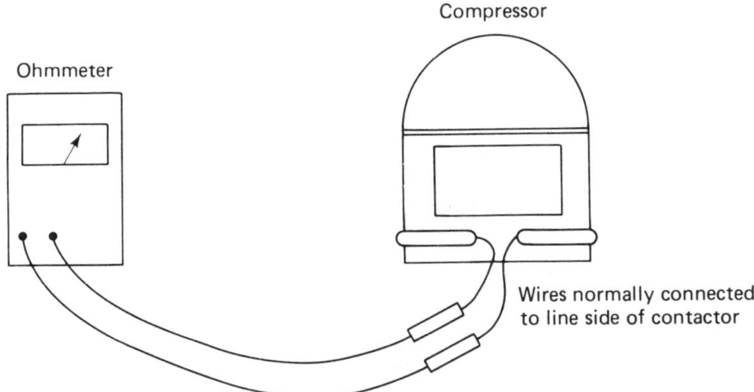

Figure 2–61 Checking resistance of crankcase heater.

Externally mounted crankcase heaters are manufactured by several manufacturers and are easily installed on most compressors. The internally mounted heaters are usually designed for a specific model of compressor. To check crankcase heaters, dampen a finger and gently touch the heater. If it is working, it will be hot to the touch. The electrical power must be on for several hours before this test can be performed. Another method of checking is to disconnect both electrical wires and check the resistance of the heater (see Figure 2-61). Crankcase heaters are generally energized continuously and are designed to prevent overheating of the compressor oil.

2.22 TWO—SPEED CONDENSER FAN THERMOSTAT

Two-speed condenser fan motors are controlled by a thermostat mounted on a return bend of the condensing coil. They sense both the outdoor and refrigerant temperatures and automatically change the speed of the fan. These controls are nonadjustable and switch the fan to low speed when the outdoor temperature is about 75° F (23.8° C) and refrigerant temperature is about 95° F (35.0° C). The fan is switched into high speed when the outdoor temperature reaches about 90° F (32.1° C) and refrigerant temperature is about 110° F (43.3° C) (see Figure 2-62).

To check these controls, check both the outdoor and refrigerant temperatures and if the fan is not operating in the proper mode, and electricity is directed in the proper circuit, the thermostat is defective and must be replaced. Be certain that electricity is supplied to the unit.

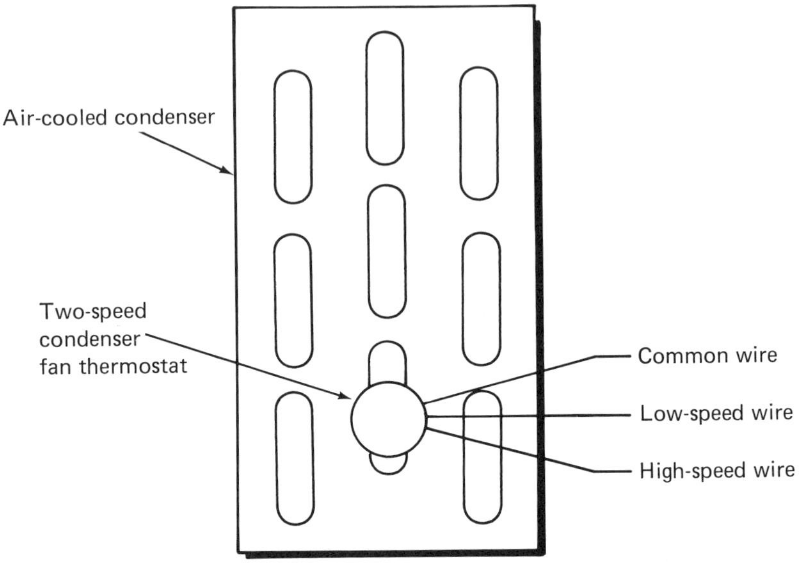

Figure 2–62 Location of two-speed condenser fan thermostat.

2.23 UNEQUALIZED PRESSURES

Unequalized pressures is a term used to indicate that the refrigerant pressures in the high and the low sides of the system are not close enough together to allow a PSC motor to start the compressor. When there is a great difference between these two pressures, the starting load is too great for the motor to start. Less starting torque is required as the difference between these two pressures become less; that is, with zero pressure differential between the high and the low side, a minimum of starting torque is required. There are two remedies to this situation. One is to allow the system to sit idle for a longer period of time using a timing device (see Section 2.24). The second is to install a hard-start kit (see Section 2.25).

2.24 COMPRESSOR TIMER

A compressor timer is a device used to prevent the compressor from cycling too soon after shutdown. This control prevents the starting of the compressor until the pressures have equalized to a safe starting point. This procedure helps to prolong the life of the compressor (see Figure 2-63).

At the end of an on-cycle the timer goes through a three-minute period when the compressor cannot restart. After the three minutes have passed, the compressor can restart and operate normally. In most cases the timer is used on units that do not have a hard-start kit installed.

To check this control, after the proper time period has passed with the system demanding operation, jumper the terminals R1 to Y1 and R2 to Y2. If the timer is defective the compressor should start. If it does, replace the timer with a proper replacement.

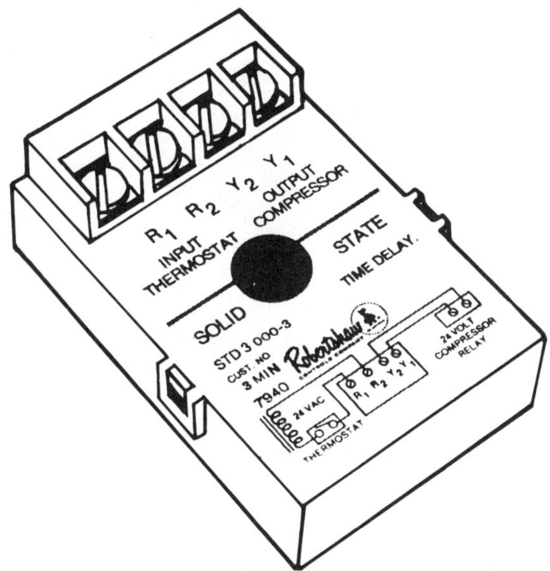

Figure 2–63 Compressor timer (Courtesy of Robertshaw Controls Co., Uni-Line Division).

2.25 HARD-START KIT

Hard-start kits are designed for use when conditions are encountered that prevent the PSC motor from starting normally. Such conditions are when the electrical power fluctuates or is too low to provide the necessary power for the proper starting of a compressor motor. Another such condition occurs when a PSC motor is used on a system that requires rapid cycle operation, less pressure equalization. Hard-start kits are designed to convert PSC motors to CSCR (capacitor-start capacitor-run) motors. Hard-start kits consist of the proper starting relay and the proper starting capacitor, along with the necessary wiring to install the kit on the unit. The individual components may also be combined to make up a hard-start kit. To install a hard-start kit, turn off the electrical power to the unit and complete the connections as shown in Figure 2-64.

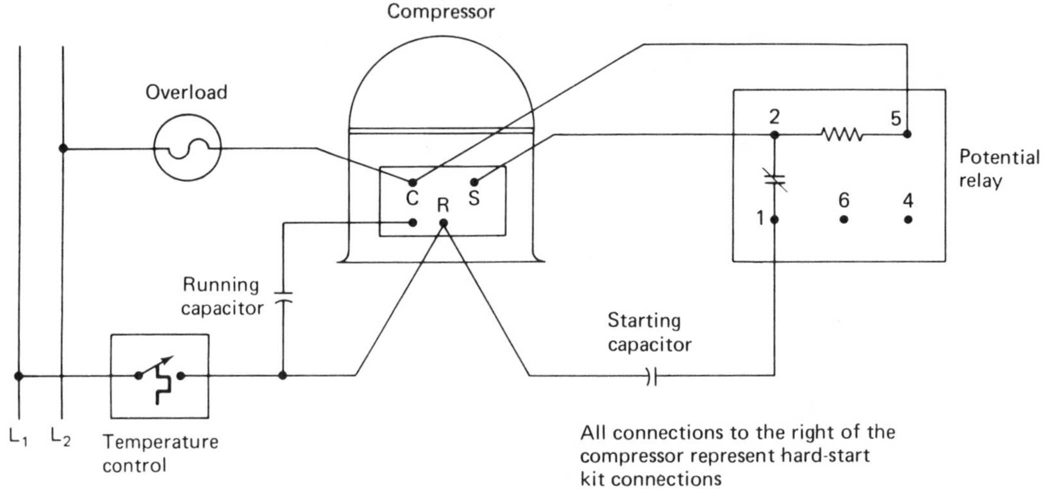

Figure 2–64 Diagram showing hard-start kit connections.

2.26 HIGH DISCHARGE PRESSURE

A high discharge pressure can be the result of one or a combination of things. It is a condition that causes an overload on the motor and decreases the efficiency of the compressor and the refrigeration system. The most common causes are: (1) compressor discharge service valve front-seated; (2) lack of cooling air; (3) lack of cooling water; (4) an overcharge of refrigerant; (5) noncondensable gases; or (6) a combination of these things.

2.26.1 Compressor Discharge Service Valve

The compressor discharge service valve when front-seated will reduce or completely stop the flow of refrigerant from the compressor. Caution should be exercised to prevent this condition, because damage to the compressor or motor is likely. The rapid buildup of pressure within the compressor cylinder head is tremendous and increases rapidly as the piston completes its upward stroke. Never front-seat the compressor discharge valve while the compressor is running or start the compressor with the valve front-seated.

2.26.2 Lack of Cooling Air

The lack of cooling air over an air-cooled condenser will cause the discharge pressure to increase. This is because the higher pressure is required to condense the vapor to a liquid. The higher refrigerant temperature also reduces the unit efficiency because of the increased flash-gas, as well as reducing the compressor efficiency. This condition is generally caused by a dirty condenser coil, loose condenser fan belt, or bad condenser fan motor bearings.

A dirty condenser can be cleaned by using a garden hose with a high-pressure nozzle. The water should be forced through the condenser from both sides. Be sure to prevent water from entering the fan motor, which might cause an electrical short. This can be done by wrapping a piece of sheet plastic around the motor (see Figure 2-65). Be sure to remove the plastic before starting the unit. Turn the unit off during this procedure.

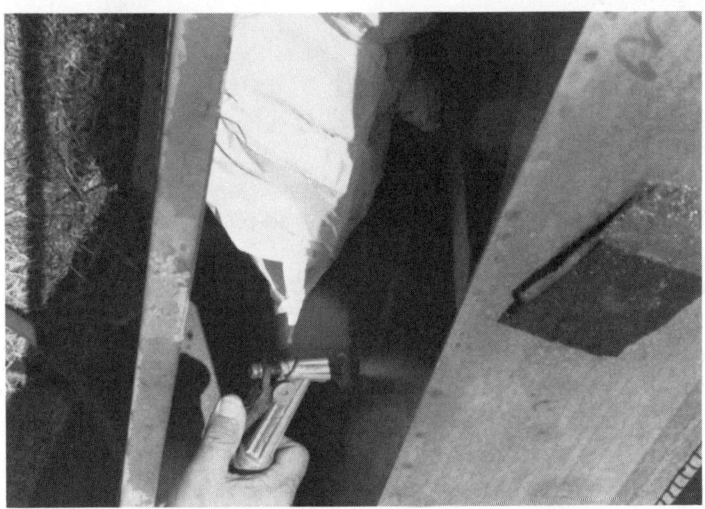

Figure 2–65 Cleaning air cooled condenser with a garden hose and pressure nozzle.

A loose or broken fan belt will prevent the blower from moving air to cool the condenser. This condition is generally obvious and is easily corrected. To adjust the belt tension, turn the adjustment until the belt can be flexed about 1 in. with one finger using moderate pressure (see Figure 2-66). If the belt is frayed, has wear grooves on the sides, or has become hard, it should

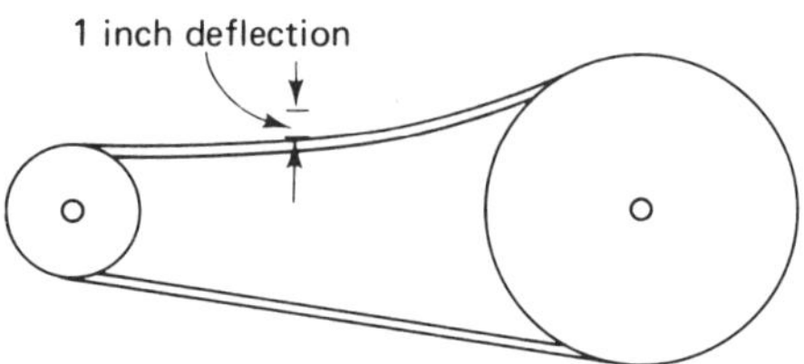

Figure 2–66 Proper belt tension adjustments.

be replaced. Be sure to use the proper size belt. A belt that is too narrow will ride the bottom of the pulley (see Figure 2-67). It will slip, causing decreased efficiency. A belt that is too wide will ride high in the pulley and will not maintain the desired efficiency and will probably overload the fan motor.

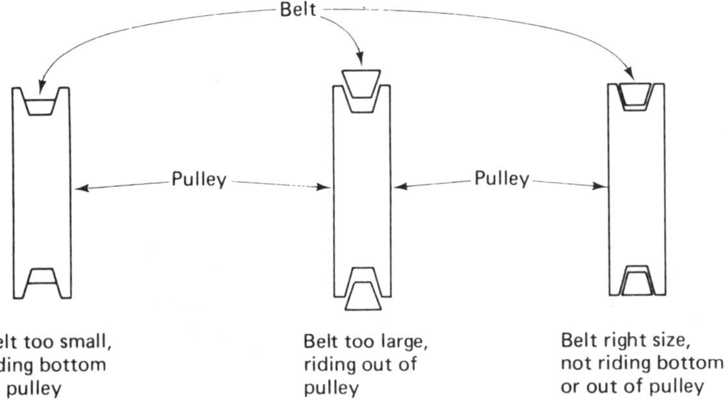

Figure 2–67 Comparison of belt and pulley fitting.

Bad fan motor bearings will cause the motor to overheat and cut out on the overload. When the condenser fan motor stops, the condenser overheats, and the compressor will cut out on high pressure. To check for bad bearings, stop the unit, remove the belt, and move the motor shaft from side to side, not end to end. Any movement of the shaft in a sideways direction indicates bad bearings. The bearings must be replaced or the motor replaced, depending on the motor size.

2.26.3 Lack of Cooling Water

The lack of cooling water on a water-cooled condenser will cause the discharge pressure to increase because of the higher pressure required to condense the vapor to a liquid. The higher refrigerant temperature also reduces the unit efficiency because of the increased amount of flash-gas, as well as reducing the compressor efficiency. This condition is generally caused by plugged water strainers, pumps, or spray nozzles.

When this condition is suspected, check the temperature rise of the water as it passes through the condenser. This temperature rise should not be more than 10° F (5.56° C). If the temperature rise is greater than 10°, a strainer, pump, or spray nozzle is stopped up or there is not enough water in the cooling tower water sump. The necessary steps must be taken to relieve this condition.

A temperature rise of less than 10° F (5.56° C) indicates that the condenser is scaled and must be cleaned. There are several commercial cleaners available for cleaning (acidizing) water-cooled condensers. The amount used is recommended by the manufacturer of the cleaner. Caution should be exercised to prevent damage to the equipment, personnel, and the surrounding vegetation when acidizing a unit. The most common method of acidizing a unit is to be sure the strainers, pump, and spray nozzles are clean. Then dump the cleaner into the tower sump and check the mixture with pH strips, adding cleaner until the desired pH is indicated, and allow the unit to run until all the scale has been removed. The cleaner and sediment must be completely removed from the system. To do this, drain and flush the tower, condenser, and water lines with fresh water. Then add a neutralizer, allowing the system to run until the neutralizer has had time to neutralize the cleaner; then drain and refill the system with fresh water. Never leave the cleaner in the system because of possible damage that may be done to the equipment.

Another method is to disconnect the water lines from the condenser and circulate the cleaner through the condenser with a separate acid pump. This method is usually expensive and therefore not popular.

2.26.4 Water-cooled Condensers

These are sometimes equipped with a valve that controls the amount of water passing through the condenser (see Figure 2-68). These valves eventually become clogged with minerals, causing them to become inoperative. When this occurs, the valve must be removed and either replaced or repaired. Be sure to relieve the pressure in that part of the system where the pressure attachment is connected. When the valve is repaired or replaced, adjust the valve to provide the proper discharge pressure (see Figure 2-69).

2.26.5 Water Used to Cool Water-cooled Condensers

The water must have a temperature low enough to cause the refrigerant vapor to condense at the pressures encountered in the normal operating cycle. Cooling towers are used to cool the water (see Figure 2-70).

The spray nozzles cause the water to atomize and mix with the air. This

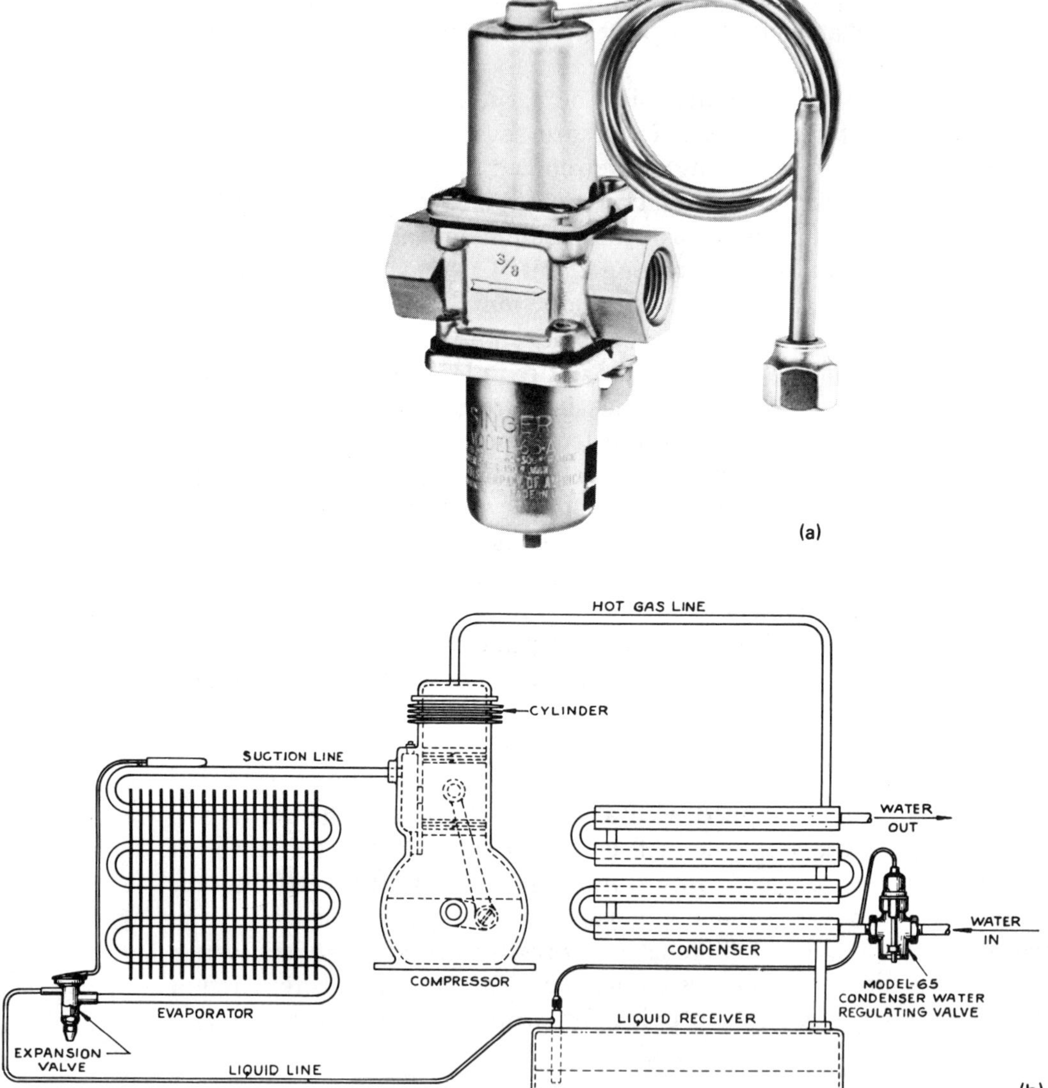

Figure 2–68 (a) Water flow control valve (Courtesy of Singer Controls Division) and (b) location of water flow control valve.

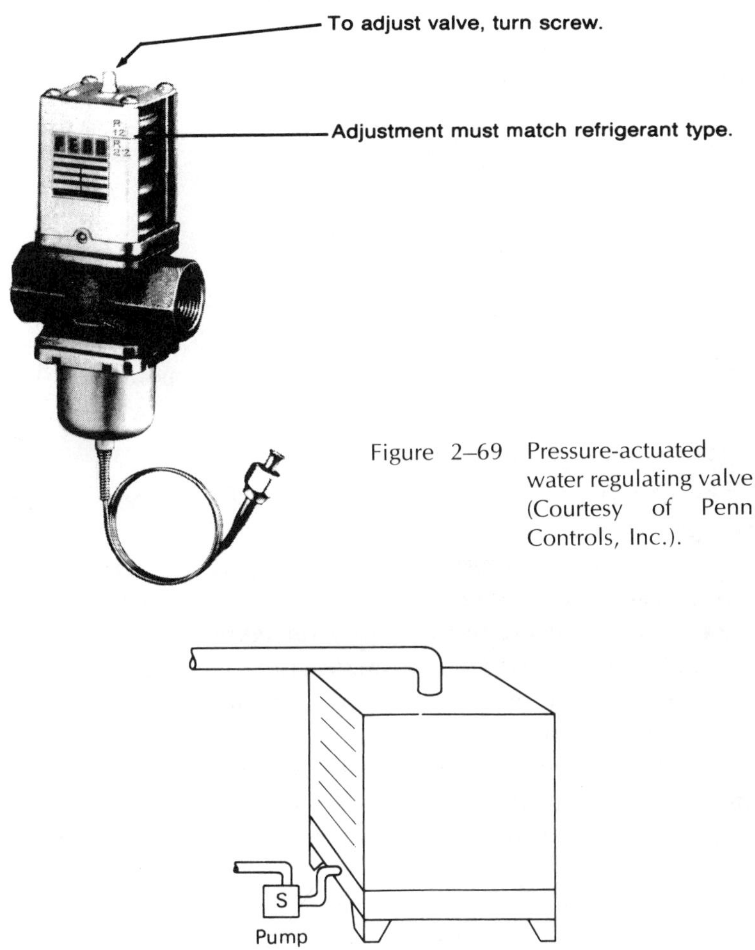

Figure 2–69 Pressure-actuated water regulating valve (Courtesy of Penn Controls, Inc.).

Figure 2–70 Cooling tower.

mixing with the air causes some of the water to evaporate, lowering its temperature. If the spray nozzles do not atomize the water, it will not be cooled sufficiently. To determine whether or not the cooling tower is functioning properly, measure the water temperature as it leaves the tower sump and as it leaves the spray nozzle. The difference in these two temperature readings should be a minimum of 10°F (5.56°C) see Figure 2-71.

Should the spray nozzles become clogged, the water will not be cooled sufficiently. Be sure that a full cone of water is coming from each nozzle. It may be necessary to clean the pump and strainers so that the required amount of water can be circulated through the system.

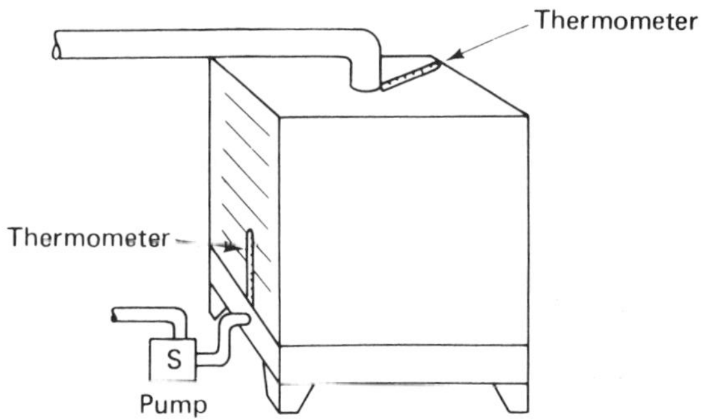

Figure 2–71 Checking cooling tower water temperatures.

2.26.6 Refrigerant Overcharge

An overcharge refrigerant will cause a high discharge pressure because the extra refrigerant will take up space in the condenser that is needed to condense the vapor to a liquid. An overcharge of refrigerant will cause at least one-half the condenser tubes to be cooler than the remainder. The cool tubes are the ones that are full of liquid refrigerants (see Figure 2-72).

When a capillary tube is used, the suction pressure will also be higher than normal. The suction line will be cooler than normal and may be frosted over, depending on the amount of overcharge.

The excess of refrigerant must be purged from the system. When purging, allow only a small amount of refrigerant to escape at a time. This procedure is recommended to prevent purging too much refrigerant from the system,

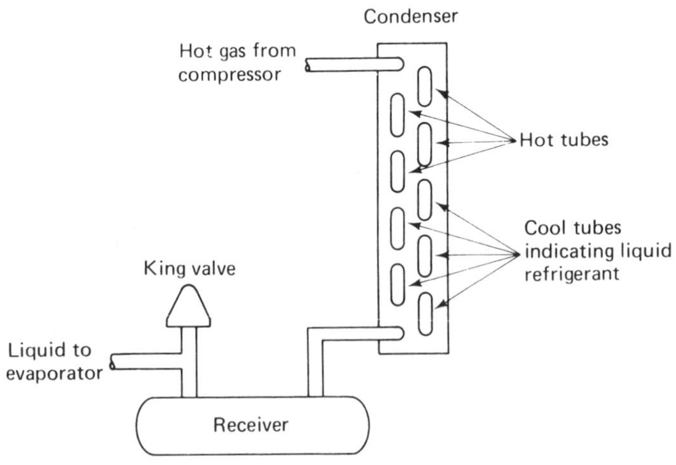

Figure 2–72 Checking liquid level in condenser.

which would require that refrigerant be added back into the unit. Also, it prevents the removal of lubricating oil from the compressor crankcase.

2.26.7 Noncondensables

The presence of noncondensables in a system will cause a high discharge pressure because they will not condense under the normal operating pressures encountered in refrigeration systems and will take up space needed by the refrigerant in the condenser. To determinc if noncondensables are present in a system, pump the system down; that is, pump all the refrigerant into the receiver or condenser by front-seating the liquid line service (king) valve (see Figure 2-73).

Operate the compressor until the low side has been pumped down to approximately 5 psig (135.34 kPa). Stop the compressor and allow it to stand idle until it has cooled to the ambient temperature. Compare the idle discharge pressure to the pressure indicated on a pressure-temperature chart at that ambient temperature. If the discharge pressure is higher than that indicated by the chart, slowly purge the noncondensables from the highest point in the system. Purge the air slowly to prevent purging an excess of refrigerant from the system. If the system contains only a small charge of refrigerant, it will probably be better to purge the entire charge, evacuate the system, and add a complete new charge. Noncondensables should be kept out of the system because they not only reduce the system efficiency but they also contain moisture, which is extremely harmful to the refrigeration system.

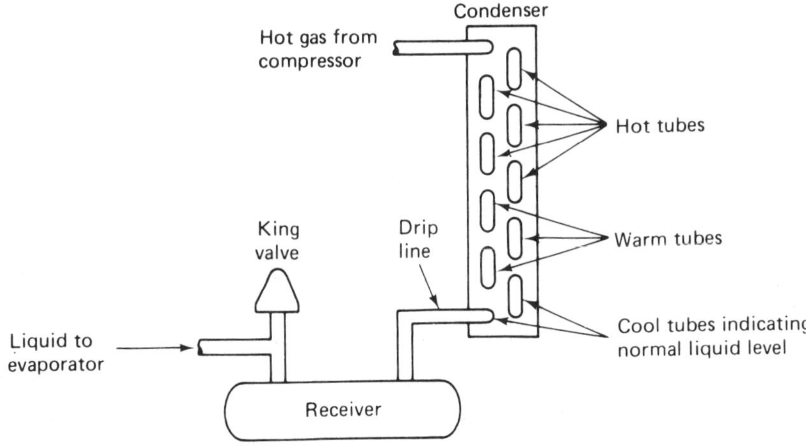

Figure 2–73 Location of king valve.

2.27 LOW SUCTION PRESSURE

A suction pressure that is lower than normal may be due to any or all of the following reasons: (1) shortage of refrigerant; (2) dirty air filter; (3) dirty evaporator; (4) dirty blower; (5) loose fan belt; (6) iced-over evaporator; (7) TXV (thermostatic expansion valve) superheat set too low; (8) AXV (automatic expansion valve) set too low; or (9) restriction in the refrigeration system. The problem or problems causing low suction pressure should be found and corrected because the compressor may pump all the lubricating oil out of its crankcase and into the system, resulting in possible damage to the compressor as well as inefficient operation of the unit due to an oil logged evaporator.

2.27.1 Refrigerant Shortage

A shortage of refrigerant is usually due to a leak that has developed in the refrigerant circuit. The leak should be found and repaired and the system recharged with a proper charge of refrigerant. If the leak is not found and repaired, the refrigerant will escape from the system, requiring more service, A refrigerant leak is generally indicated by oil on the place where the leak has occured. Many times a visual inspection will locate the problem area (see Figure 2-74).

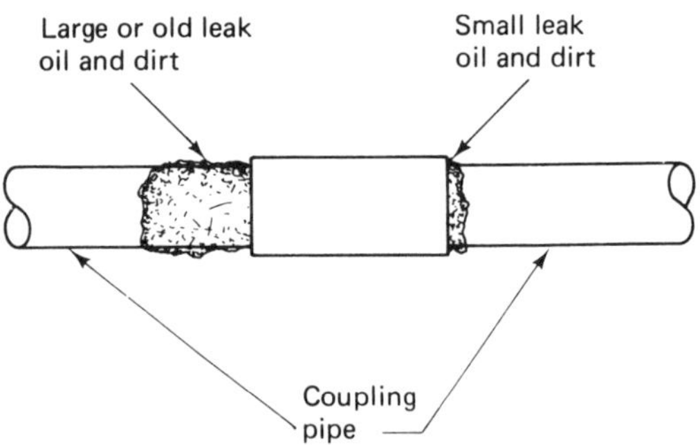

Figure 2–74 Oil around refrigerant leak.

If the problem is not found by visual inspection, an electronic or halide torch leak detector should be used. Either of these will pinpoint a leak under these conditions. Two conditions that make leak detection difficult with these tools are when the wind is blowing outdoors or in an enclosed space where the refrigerant concentration is high. When these conditions are encountered, a soap and water solution or one of the liquid plastic gas leak detectors must be used. The leak may be found by applying the solution to the suspected area and watching for bubbles to appear (see Figure 2-75).

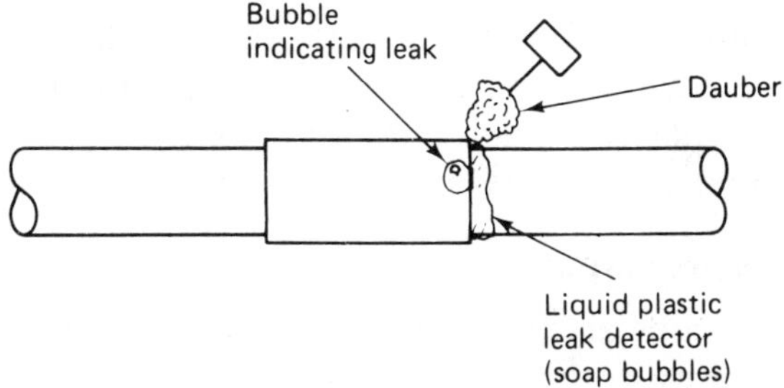

Figure 2–75 Leak testing with liquid plastic leak detector (soap bubbles).

The bubbles will appear in only a few seconds. When repairing leaks that require heating the tubing, be sure to relieve any pressure from inside the tubing that is being heated. This will prevent a blowout, which can result in damage to the equipment and injury to the service technician.

2.27.2 Dirty Air Filters

These are a frequent cause of low suction pressure in air conditioning systems. Dirty air filters restrict the flow of air over the evaporator, resulting in a low load on the refrigeration system. The air filter is located on the air inlet side of the blower (see Figure 2-76).

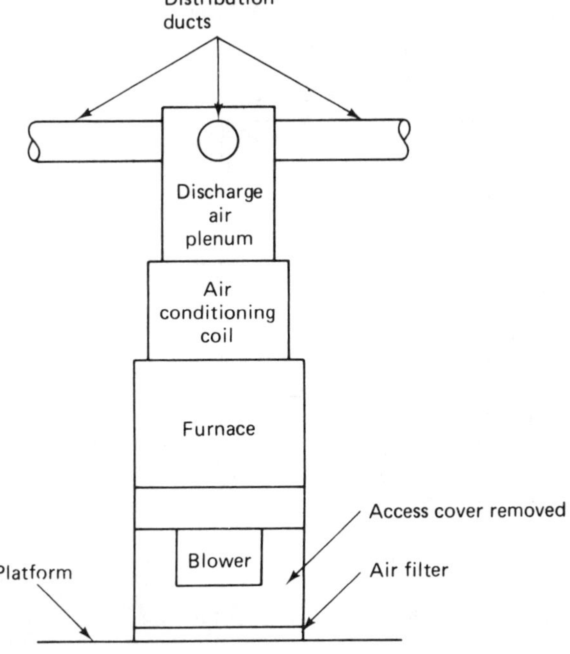

Figure 2–76 Air filter location.

The filters should be changed or cleaned on a regular basis to ensure peak efficiency from the unit. The throwaway-type filters should not be cleaned. They should be replaced. When cleanable filters are used, they should be coated with a filter coater after cleaning to increase their effectiveness.

2.27.3 Dirty Evaporator Coils

Dirty evaporator coils are often the cause of low suction pressures in air conditioning systems. A dirty evaporator restricts the air flow through the unit and the insulating valve of the dirt prevents the proper transfer of heat to the refrigerant. A dirty evaporator coil is the result of a dirty air filter or one that does not fit properly. When an air filter does not fit properly, the dust-laden air will bypass it, and the dust will stick to the evaporator coil fins (see Figure 2-77).

Precautions must be taken to ensure a proper fit of the air filter, and the evaporator must be cleaned before the system will function properly. Many times the evaporator will need to be removed and be steam-cleaned. Be sure to prevent moisture entering the refrigerant system during the cleaning process.

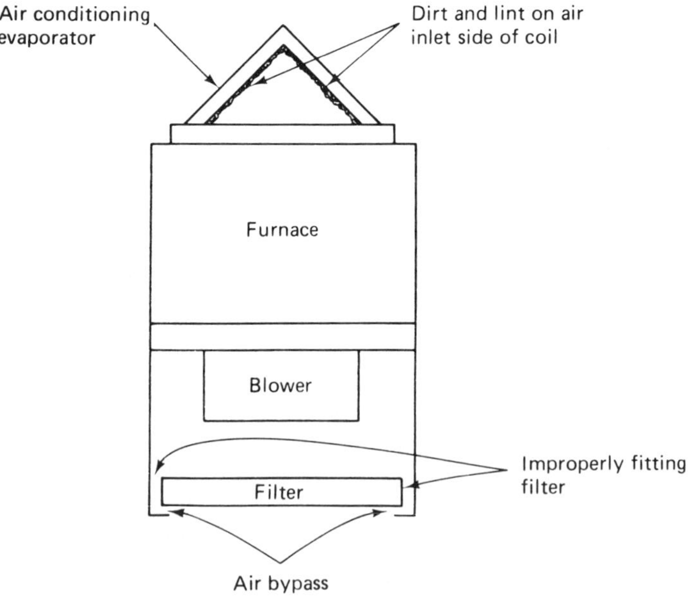

Figure 2–77 Dirty coil because of filter air bypass.

2.27.4 Dirty Blowers

A dirty blower will not deliver enough air to place the proper load on the evaporator, resulting in a low suction pressure. A dirty blower is the result of improperly fitting air filters or a unit that has been operated with an excessively dirty filter. The blower must be removed and the vanes cleaned so that only the metal can be seen. The filter must be cleaned, or replaced, and any air bypass must be prevented. Otherwise, the problem will reoccur. When steam-cleaning or washing the blower with water, take precautions to prevent moisture entering the electric blower motor.

2.27.5 Loose or Broken Fan Belts

A loose or broken fan belt will prevent proper operation of the blower and reduce the flow of air over the evaporator, resulting in a low load or no load on the evaporator and a low suction pressure. A loose belt that is in good condition can be properly adjusted and the unit put back in service. A broken or frayed belt or one that has become hard on the edges must be replaced and properly adjusted. To properly adjust a belt, tighten the adjustment screw until the belt can be flexed about 1 in. with one finger using moderate pressure (see Figure 2-66). Be sure to use the proper sized belt. A belt that is too narrow will ride the bottom of the pulley and slip, causing decreased efficiency (see Figure 2-67). A belt that is too wide will ride too high in the pulley and will not maintain the desired efficiency and will possibly overload the motor.

2.27.6 Iced or Frosted Evaporator

An iced or frosted evaporator will cause a low suction pressure because the insulating action of the ice or frost will prevent proper heat transfer from the air to the refrigerant. An iced evaporator is usually caused by an insufficient load on the refrigeration system or a shortage of refrigerant. A reduced load can be the result of (1) a dirty filter, (2) a dirty evaporator coil, (3) a dirty blower, or (4) a loose or broken fan belt. A shortage of refrigerant is due to a refrigerant leak. Any of these conditions must be corrected and the coil de-iced before the system will operate satisfactorily. De-icing can be accomplished by turning on the fan and turning off the compressor. Allow the fan to operate until the ice has been melted. A faster method is to apply a small amount of heat to the ice. Caution should be exercised to prevent overheating of the evaporator or the fins.

2.28 THERMOSTATIC EXPANSION VALVES

Thermostatic expansion valves are the most common type of refrigerant flow control devices used on air conditioning units of 5-ton capacity and above. They operate in response to both pressure and temperature inside the evaporator. The proper superheat adjustment of these valves will have a tremendous effect on the efficiency of the equipment.

A thermostatic expansion valve that is adjusted for too great a superheat or an improperly located feeler bulb will result in a low suction pressure. If the superheat is found to be improperly adjusted, it may be adjusted by the following procedure:

1. Measure the temperature of the suction line at the point where the feeler bulb is clamped.
2. Determine the suction pressure that exists in the suction line at the bulb location by either of the following methods:
   a. If the valve is externally equalized, a pressure gauge in the external equalizer line will indicate the correct pressure directly and accurately.
   b. Read the gauge pressure at the suction service valve of the compressor. To the reading add the estimated pressure drop through the suction line between the bulb location and the compressor service valve. The sum of the gauge reading and the estimated pressure drop will equal the approximate line pressure at the bulb.
3. Convert the pressure reading obtained in Step 2(a) or Step 2(b) to the saturated evaporator temperature by using the pressure-temperature chart as shown in Table 2-1.
4. Subtract the two temperatures obtained in Step 1 and Step 3. The difference is the superheat setting of the expansion valve.

Figure 2-78 illustrates a typical example of superheat measurement on an air conditioning system using Refrigerant 12 as the refrigerant.

The temperature of the suction line at the bulb location is read at 51°F (10.6°C). The suction pressure at the compressor suction service valve is 35 psig (241.316 kPa), and the estimated pressure drop is 2 psig (12.790 kPa). The total suction pressure is 35 psig + 2 psig = 37 psig (255.106 kPa). This is equivalent to a 40°F (4.4°C) saturation temperature. Then, 40°F subtracted from 51°F equals 11°F (6.2°C) superheat setting.

Table 2–1 Pressure-Temperature Chart
(2–1 TS, Courtesy of Alco Controls Division).

BOLD FIGURES = INCHES MERCURY VACUUM LIGHT FIGURES = PSIG

°F	R-12	R-13	R-22	R-500	R-502	R-717 Ammonia	°F	R-12	R-13	R-22	R-500	R-502	R-717 Ammonia
−100	**27.0**	7.5	**25.0**	–	**23.3**	**27.4**	16	18.4	211.9	38.7	24.2	47.8	29.4
−95	**26.4**	10.9	**24.1**	–	**22.1**	**26.8**	18	19.7	218.8	40.9	25.7	50.1	31.4
−90	**25.7**	14.2	**23.0**	–	**20.7**	**26.1**	20	21.0	225.7	43.0	27.3	52.5	33.5
−85	**25.0**	18.2	**21.7**	–	**19.0**	**25.3**	22	22.4	233.0	45.3	29.0	55.0	35.7
−80	**24.1**	22.2	**20.2**	–	**17.1**	**24.3**	24	23.9	240.3	47.6	30.7	57.5	37.9
−75	**23.0**	27.1	**18.5**	–	**15.0**	**23.2**	26	25.4	247.8	49.9	32.5	60.1	40.2
−70	**21.8**	32.0	**16.6**	–	**12.6**	**21.9**	28	26.9	255.5	52.4	34.3	62.8	42.6
−65	**20.5**	37.7	**14.4**	–	**10.0**	**20.4**	30	28.5	263.2	54.9	36.1	65.4	45.0
−60	**19.0**	43.5	**12.0**	–	**7.0**	**18.6**	32	30.1	271.3	57.5	38.0	68.3	47.6
−55	**17.3**	50.0	**9.2**	–	**3.6**	**16.6**	34	31.7	279.5	60.1	40.0	71.2	50.2
−50	**15.4**	57.0	**6.2**	–	0.0	**14.3**	36	33.4	287.8	62.8	42.0	74.1	52.9
−45	**13.3**	64.6	**2.7**	–	2.1	**11.7**	38	35.2	296.3	65.6	44.1	77.2	55.7
−40	**11.0**	72.7	0.5	**7.9**	4.3	**8.7**	40	37.0	304.9	68.5	46.2	80.2	58.6
−35	**8.4**	81.5	2.6	**4.8**	6.7	**5.4**	45	41.7	327.5	76.0	51.9	88.3	66.3
−30	**5.5**	91.0	4.9	**1.4**	9.4	**1.6**	50	46.7	351.2	84.0	57.8	96.9	74.5
−28	**4.3**	94.9	5.9	0.0	10.6	0.0	55	52.0	376.1	92.6	64.2	106.0	83.4
−26	**3.0**	98.9	6.9	0.7	11.7	0.8	60	57.7	402.3	101.6	71.0	115.6	92.9
−24	**1.6**	103.0	7.9	1.5	13.0	1.7	65	63.8	429.8	111.2	78.2	125.8	103.1
−22	**0.3**	107.3	9.0	2.3	14.2	2.6	70	70.2	458.7	121.4	85.8	136.6	114 1
−20	0.6	111.7	10.1	3.1	15.5	3.6	75	77.0	489.0	132.2	93.9	148.0	125.8
−18	1.3	116.2	11.3	4.0	16.9	4.6	80	84.2	520.8	143.6	102.5	159.9	138.3
−16	2.1	120.8	12.5	4.9	18.3	5.6	85	91.8	–	155.7	111.5	172.5	151.7
−14	2.8	125.7	13.8	5.8	19.7	6.7	90	99.8	–	168.4	121.2	185.8	165.9
−12	3.7	130.5	15.1	6.8	21.3	7.9	95	108.3	–	181.8	131.2	199.7	181.1
−10	4.5	135.4	16.5	7.8	22.8	9.0	100	117.2	–	195.9	141.9	214.4	197.2
−8	5.4	140.5	17.9	8.8	24.4	10.3	105	126.6	–	210.8	153.1	229.7	214.2
−6	6.3	145.7	19.3	9.9	26.0	11.6	110	136.4	–	226.4	164.9	245.8	232.3
−4	7.2	151.1	20.8	11.0	27.7	12.9	115	146.8	–	242.7	177.3	262.6	251.5
−2	8.2	156.5	22.4	12.1	29.5	14.3	120	157.7	–	259.9	190.3	280.3	271.7
0	9.1	162.1	24.0	13.3	31.2	15.7	125	169.1	–	277.9	203.9	298.7	293.1
2	10.2	167.9	25.6	14.5	33.1	17.2	130	181.0	–	296.8	218.2	318.0	315.0
4	11.2	173.7	27.3	15.7	35.0	18.8	135	193.5	–	316.6	233.2	338.1	335.0
6	12.3	179.8	29.1	17.0	37.0	20.4	140	206.6	–	337.3	248.8	359.1	365.0
8	13.5	185.9	30.9	18.4	39.1	22.1	145	220.6	–	358.9	265.2	381.1	390.0
10	14.6	192.1	32.8	19.8	41.1	23.8	150	234.6	–	381.5	282.3	403.9	420.0
12	15.8	198.6	34.7	21.2	43.3	25.6	155	249.9	–	405.2	300.1	427.8	450.0
14	17.1	205.2	36.7	22.7	45.5	27.5	160	265.12	–	429.8	318.7	452.6	490.0

* Courtesy of Alco Controls Division

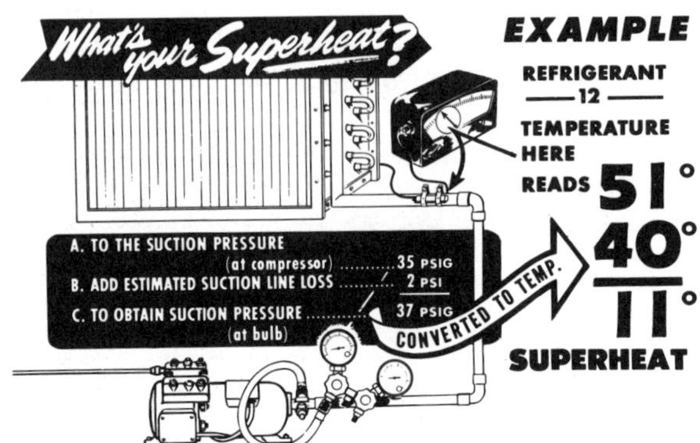

Figure 2–78 Determination of superheat (Courtesy of Sporlan Valve Co.).

Notice that subtracting the difference between the temperature at the evaporator inlet and the temperature at the evaporator outlet is not an accurate measurement of the superheat. This method is not recommended because any evaporator pressure drop will result in an erroneous superheat indication.

The valve should be adjusted to provide the desired superheat setting. To determind if the expansion valve is operating, remove the remote bulb and warm it with the hand while observing the suction pressure. If the pressure increases, the valve needs adjusting.

2.28.1 Feeler Bulb on a Thermostatic Expansion Valve

This is one of the major forces controlling the valve. If the bulb is located in a cold location, it will cause the valve to starve the evaporator, causing a lower than normal suction pressure. The location of the feeler bulb is extremely important and in some cases determine the success or failure of the refrigeration plant. For satisfactory expansion valve control, good thermal contact between the bulb and the suction line is essential. The bulb should be securely fastened with two bulb straps to a clean, straight section of the suction line.

Installation of the bulb to a horizontal run of suction line is preferred. If a vertical installation cannot be avoided, the bulb should be mounted so that capillary tubing comes from the top (see Figure 2-79). To install, clean the suction line thoroughly before clamping the remote bulb in place. When a steel suction line is used, it is advisable to paint the line with aluminum paint to minimize future corrosion and to eliminate faulty remote bulb contact with the line. On lines under 7/8 in. (22-225 mm) OD (outside diameter), the remote bulb may be installed on top of the line. On 7/8 in. OD and larger,

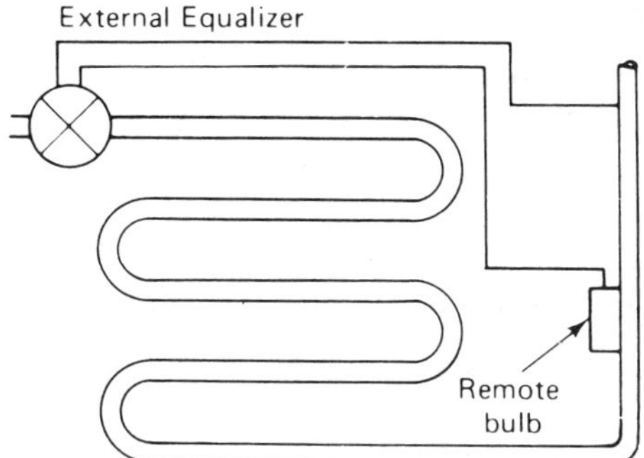

Figure 2–79 Remote bulb installation on vertical tubing.

the remote bulb should be installed at about the 4 or 8 o'clock position (see Figure 2-80).

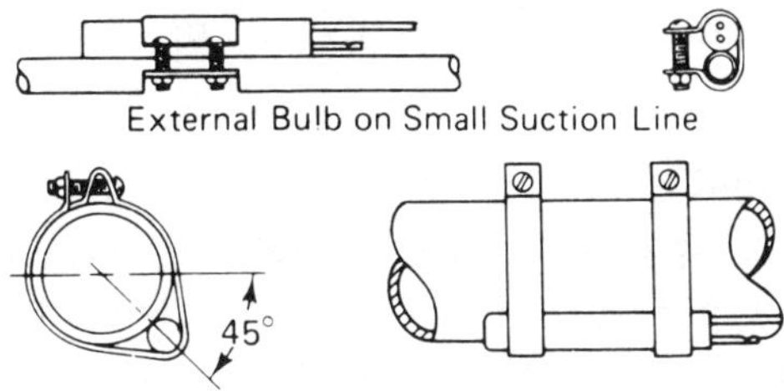

Figure 2–80 Remote bulb installation on horizontal tubing. (Courtesy of Alco Controls Division, Emerson Electric Co.)

If it is necessary to protect the remote bulb from the effects of the air stream after it is clamped to the line, use a material such as sponge rubber that will not absorb water when the evaporator temperatures are above 32° F (0° C). Below 32° F, cork or similar material sealed against moisture is suggested to prevent ice collecting at the remote bulb location.

2.28.2 Thermostatic Expansion Valve Stuck Open

When a thermostatic expansion valve is stuck open, there will be an excessive amount of sweating on the suction line. The compressor crankcase will also sweat because of the liquid refrigerant being fed back to the compressor. It is best to replace a sticking expansion valve.

2.28.3 Thermostatic Expansion Valve Too Large

If the thermostatic expansion valve is too large for the system, it will not maintain a constant suction pressure. The feeler bulb will attempt to control the flow of liquid to maintain the superheat setting, but the oversized valve will admit too much liquid too rapidly. The feeler bulb sensing this liquid will close the valve, and the pressure in the evaporator will drop until the valve opens, admitting more liquid refrigerant. This hunting action will cause the suction pressure to fluctuate, which can be seen on the compound gauge. This variation is usually 10 psig to 15 psig (169.689 kPa to 204.038 kPa). When this condition occurs, either replace the complete valve or replace the valve seat and needle.

2.28.4 Thermostatic Expansion Valve Too Small

A thermostatic expansion valve that is too small cannot pass enough liquid refrigerant to refrigerate the evaporator properly. When the unit is heavily loaded, the superheat will be high and the system capacity will be low. An expansion valve that is too small will generally cause a lower than normal suction pressure. Either replace the valve or the valve seat and needle with the proper size.

2.28.5 Power Element of Thermostatic Expansion Valve

The power element of a thermostatic expansion valve contains a vapor charge. If a leak should develop in the power element assembly, the valve will tend to close, stopping the flow of refrigerant. To check a power element, use the following procedure:

1. Stop the compressor.
2. Remove the feeler bulb from the suction line and place it in ice water.
3. Start the compressor.
4. Remove the bulb from the ice water and warm it with the hand.

At the same time feel the suction line for a drop in temperature. If the liquid refrigerant floods through the valve, the power element is operating properly. Be sure not to flood liquid refrigerant back through the suction line for a long period of time because it can cause damage to the compressor.

2.29 AUTOMATIC EXPANSION VALVES

Automatic expansion valves are constant pressure-regulating devices that respond to the refrigerant pressure at the valve outlet. When they are closed off, the suction pressure may drop lower than desired. When this occurs, a minor adjustment will usually correct the problem. To adjust, install a compound gauge on the compressor suction service valve and adjust the expansion valve to obtain the desired pressure. If the valve will adjust, it is operating satisfactorily and does not need to be replaced. The final adjustment of the expansion valve should not be done until the unit has been operating from 24 to 48 hours.

2.30 RESTRICTION IN A REFRIGERANT CIRCUIT

A restriction in the refrigerant circuit will reduce the flow of refrigerant, resulting in a lower than normal suction pressure. A restriction may be in the form of a kinked line, a plugged strainer, a plugged drier, a plugged capillary tube, or ice in the orifice of the flow control device. The restriction must be removed before proper operation can be realized.

2.30.1 Kinked Lines

A kinked line occurs when the line has been bent too far resulting in a flattened place. This type of restriction can be located by visual inspection. However, if the restriction is in the liquid line, there will be a temperature difference across the kink. If the tubing is flattened enough, there will be condensation or frost on the outside of the tubing (see Figure 2-81).

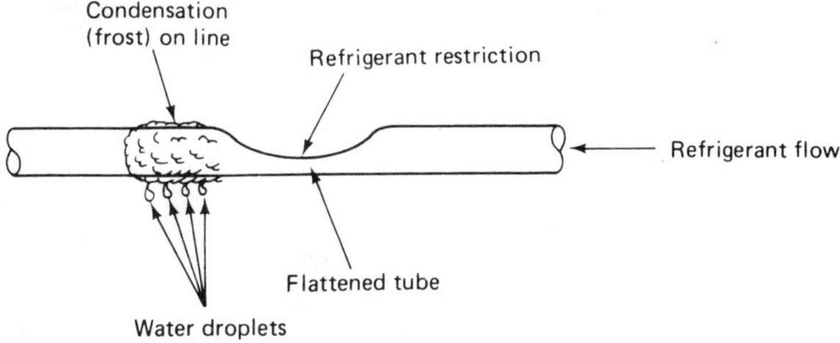

Figure 2–81 Flattened tube causing refrigerant restriction.

In cases where the tube is not flattened excessively, the flattened spot may be straightened by placing a flaring block around the tube, using the proper size hole, and tightening it down. If this fails, the kinked tubing must be removed and a new piece installed. Be sure to relieve all pressure on the tubing before attempting to make repairs. Personal injury may result when repairs are attempted while there is pressure inside the tube.

2.30.2 Strainers

The purpose of a strainer is to trap foreign particles in the refrigeration system. A plugged strainer will reduce the flow of refrigerant and may completely stop the passage of refrigerant. All expansion valves are equipped with strainers. In addition, the systems have suction line filter-driers, liquid line driers, and a compressor suction inlet strainer. All of these features are used to trap foreign materials in the system to prevent damage to the particular component being protected. A plugged strainer will develop a pressure drop across it and usually a temperature difference will exist on each side that can be felt with the bare hand. When a strainer becomes plugged, it is best to replace it if another one is readily available. If one is not available, thoroughly clean the original and replace it. Do not remove the strainer permanently. To do this will allow foreign particles to enter the protected device, causing possible damage.

2.30.3 Driers

The purpose of a drier is to remove moisture and trap foreign particles in the refrigeration system. A plugged drier will restrict or completely stop the flow of refrigerant through the system. All field built-up systems should be equipped with driers to remove any moisture and foreign particles that may accidentally enter the system during the installation process. A plugged drier can be determined by the temperature drop through it (see Figure 2-82).

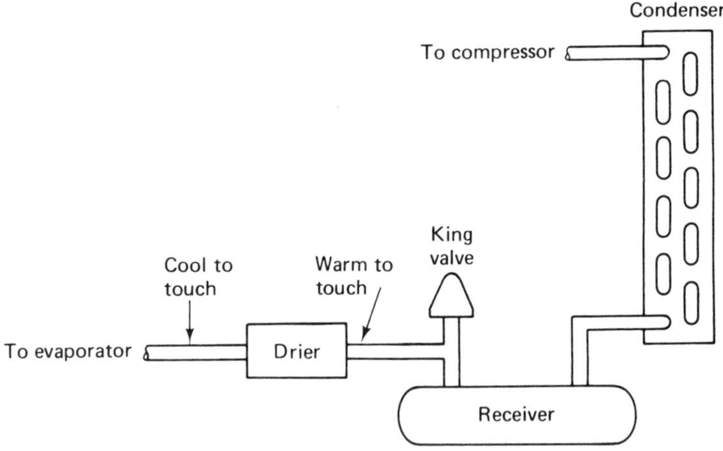

Figure 2–82 Checking a plugged drier.

A plugged drier is an indication that there are still moisture and foreign particles in the system. When removing a drier be sure to relieve all pressure inside the tube being worked on. To open a refrigeration system while it is pressurized can result in personal injury.

2.30.4 Plugged Capillary Tubes

A plugged capillary tube is caused by untrapped foreign particles in a refrigeration system and can result in a reduced or completely stopped refrigerant flow through the system. A plugged capillary tube will be indicated by a longer than usual pressure equalization time, accompanied by a loss of refrigeration. It is recommended that a plugged capillary line be replaced rather than trying to clean it out. When replacing it be sure to use the same length and inside diameter tubing as the original. To use a different size will result in an unbalanced refrigeration system and poor refrigeration. Always replace the filter-strainer along with the capillary tube. Be sure to relieve all pressure from the system before attempting repairs, or personal injury may result.

2.30.5 Ice in the Flow Control Device

Ice in the orifice of the flow control device is due to free moisture in the system. This condition generally occurs when the drier has trapped all the moisture it can absorb, and there is surplus free moisture that turns to ice in the orifice of the flow control device. This condition is usually indicated by poor operation, a lower than normal suction pressure, and low discharge pressure. To be sure that moisture is causing the problem, stop the unit and apply a cloth moistened with hot water to the outlet of the flow control device. If after a few minutes a hissing sound is heard and an increase in the low side pressure is detected, moisture is the culprit. To correct this problem a new drier must be installed. If the problem continues, it may be necessary to completely purge the refrigerant from the system, triple-evacuate the system, install oversize driers, and recharge the system with dry refrigerant.

2.31 EXCESSIVE LOAD

An excessive load condition normally occurs if the unit was undersized during the initial installation or there has been an addition to the building. The only solution to this problem is to resize the unit to better fit the application. Another condition that could cause excessive loading of the equipment occurs when the evaporator fan speed has been increased to the point of overloading the evaporator. This condition will cause an increase in the suction pressure accompanied by a warmer than normal liquid line. The obvious solution here is to reduce the fan speed while measuring the temperature drop of the air flowing through the evaporator. The temperature drop should be maintained at the manufacturer's recommendations so that the equipment will operate at peak efficiency. Usually 20° F (11.2° C) for air conditioning is recommended.

2.32 TRAPS IN LINES

Incidental traps that are left in the refrigerant lines during the installation process will trap oil and cause damage to the compressor. These traps are generally in the form of low sags in the lines. The oil will settle out of the refrigerant and collect in these low places. The refrigerant will pass through the pipe over the oil, leaving the oil in the line instead of returning it to the compressor crankcase (see Figure 2-83).

There are two ways to correct this problem. The easiest and most correct method is to install the lines properly, removing such traps. The second way, which is a bit more expensive, is installing an oil separator.

2.32.1 Oil Separators

Oil separators are installed in the compressor discharge line to collect the oil and return it to the compressor crankcase (see Figure 2-84). A proper sized

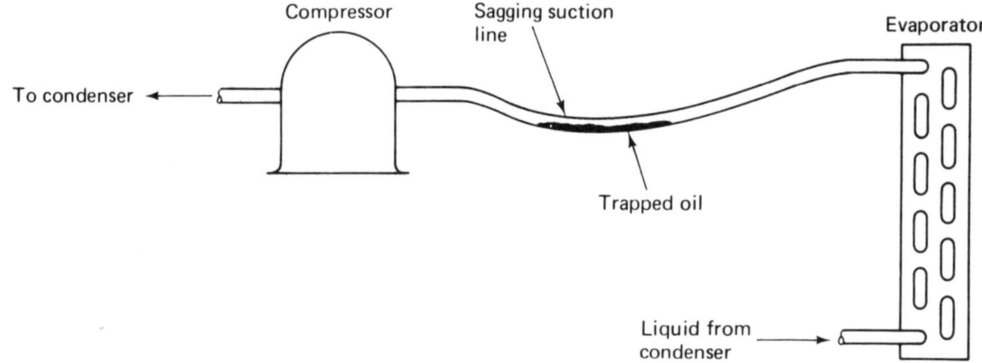

Figure 2–83 Trapped oil in suction line.

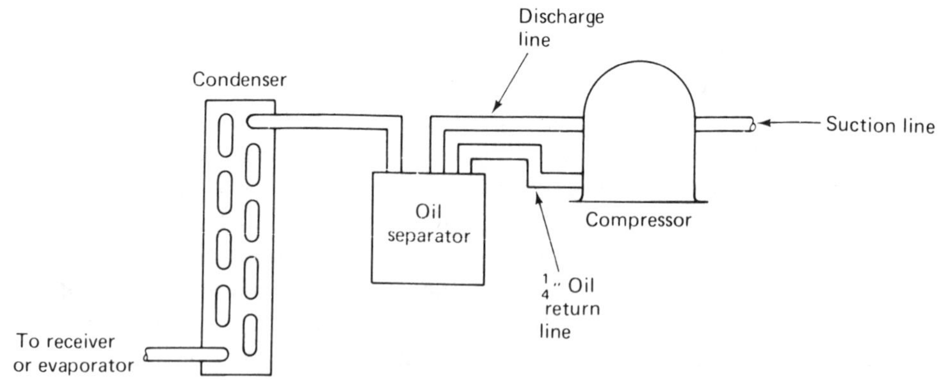

Figure 2–84 Oil separator installation.

oil separator must be used to effect proper oil return. Be sure to locate the oil separator where its temperature will not be lower than the receiver temperature during the "off" cycle. This will prevent liquid refrigerant condensing in the oil separator and being fed into the compressor crankcase.

A leaking valve in the oil separator will allow the discharge pressure to go into the compresser crankcase rather than to the condenser. This causes a lower than normal discharge pressure and a higher than normal suction pressure. The compressor crankcase will also be warmer than normal. The valve or the complete oil separator must be replaced to correct the problem.

2.32.2 Oil Return Traps

Oil return traps that are properly designed will aid in oil return on some installations where oil collects in the evaporator, a condition that frequently occurs in multiple-evaporator systems. The principal function of an oil trap is to start the oil rising in a vertical line, the refrigerant then carrying it back to the compressor. Vertical risers must be correctly sized to carry the oil to the top of the riser.

A trap that is properly constructed will have as short a horizontal run as possible. A good oil trap may be installed by connecting a street ell into a regular ell (see Figure 2-85).

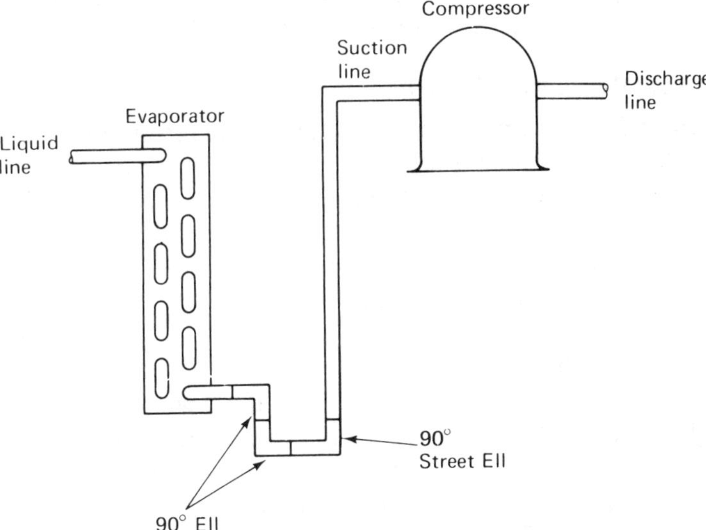

Figure 2–85 Field made oil trap made by using a street ell and a regular ell.

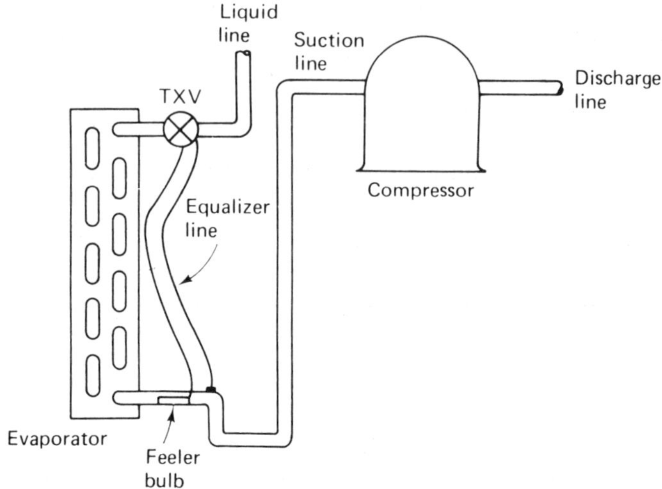

Figure 2–86 TXV bulb on horizontal line before oil trap.

On larger size piping where street ells are not available, use two regular ells with a short piece of pipe between them. When oil traps are used on systems using thermostatic expansion valves, place the feeler bulb on the horizontal line before the trap (see Figure 2-86).

2.32.3 Low Refrigerant Velocity

When the refrigerant velocity is too low in the refrigerant lines, the oil will not be returned to the compressor properly. The efficiency of the system drops, resulting in unsatisfactory operation. When this condition is found, there are two alternatives: (1) resize the lines to the proper size using tables and charts that are printed by various manufacturers; or (2) install an oil return trap as discussed in Section 2.32.2.

2.32.4 Gas-oil Ratio

Gas-oil ratio in a system results in oil flowing with the refrigerant throughout the system. The design of reciprocating compressors provides for the correct mixing of oil and refrigerant vapor. When a considerable amount of refrigerant has been lost from the system, a certain amount of oil has also been lost. The oil should be replaced along with the refrigerant. When a shortage of refrigerant results in a shortage of oil, the oil should be replaced at the ratio of 1 pt. (0.47321 g) of oil to 10 lbs. (4.5359 g) of refrigerant.

2.33 TUBING RATTLE

A tubing rattle is the result of mishandling or misuse of air conditioning equipment. Not only is this condition annoying, it is also damaging to the tubing. If let go long enough, a refrigerant leak will appear because a hole has been rubbed in the tubing. To eliminate tubing rattle, either slightly bend the tubing apart with the hands or place a cushion between the tubes. Be careful not to apply enough pressure to the lines to break or kink them.

2.34 LOOSE MOUNTING

When mountings become worn or broken, an annoying noise will result, as well as possible damage to the unit. The only proper method of correcting this situation is to replace the mounting. In cases where a bolt or nut has become loose, it can be tightened without replacement of the mounting. When springs are used as the mountings, it is usually best to replace all the springs, not just the broken one, because the rest are probably weak and may break in the near future.

2.35 COMPRESSOR DRIVE COUPLINGS

Compressor drive couplings are used on large-size compressors that are driven directly from the end of the motor shaft. A coupling is used to connect the two shafts (see Figure 2-87). There is a cushion included in the coupling to absorb the shock developed by the compressor pumping. These two shafts must be properly aligned or a vibration will be present that will cause excessive wear on the bearings and shaft seal. To align this type of coupling, loosen the hold-down bolts on either the motor or the

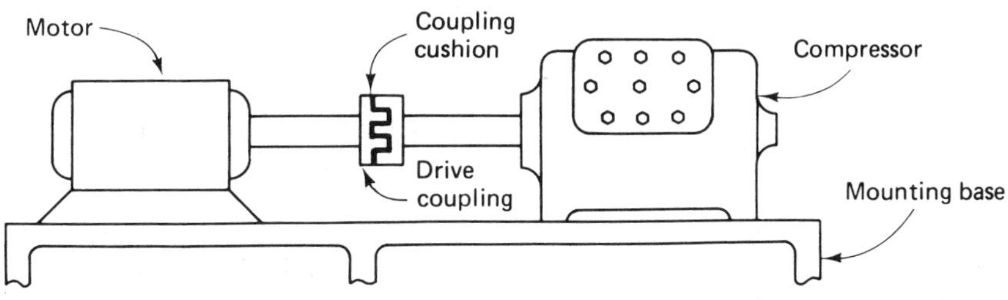

Figure 2–87 Direct drive compressor-motor coupling location.

compressor, or both. Place a straight edge on both halves of the coupling (see Figure 2-88). This must be a metal straight edge that is in good condition. Move the loosened component around until the straight edge is touching the complete face of both halves of the coupling in at least three different places around the coupling. Tighten the loosened component hold-down bolts and recheck the alignment. Then start the unit. If vibration is still present, repeat the above procedure. If the vibration still persists, replace the coupling and align it.

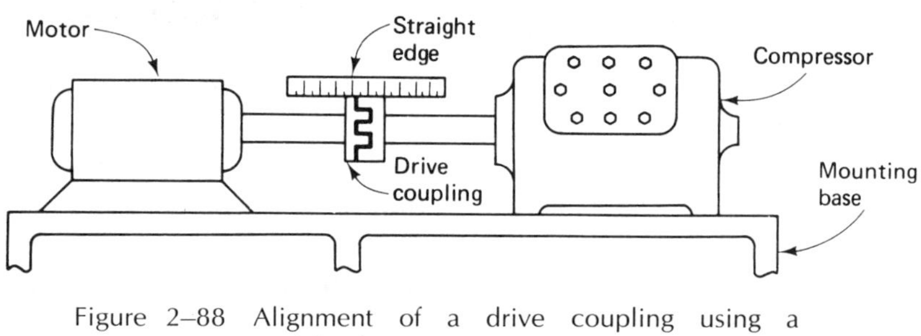

Figure 2–88 Alignment of a drive coupling using a straightedge.

2.36 EXCESSIVE PRESSURE DROP IN EVAPORATOR

An excessive pressure drop through the evaporator is generally due to a restriction in the expansion valve or in the external equalizer line. The first step in correcting the problem is to check the operation of the expansion valve as outlined in Section 2.28. If the trouble is not found there, check the external equalizer line to see that it is not restricted. If a restriction is found, it may be removed by connecting a drum of the same type refrigerant as that in the system and blowing out the restriction. Complete replacement of the equalizer line may be required. The unit should function properly when the restriction is removed.

2.37 EVAPORATOR TOO SMALL

In a properly designed and installed system, the evaporator will be of the correct size. However, sometimes a service technician will find a defective coil and will replace it with one that is too small. The only step required is to replace the evaporator with one of the proper size. An evaporator that is too small will cause a lower than normal suction pressure with probable frosting of the evaporator. As a result, it will be almost impossible to adjust the expansion valve to the required superheat setting, and the temperature of the space will be higher than desired. Therefore, the evaporator should be replaced.

2.38 LIQUID FLASHING IN LIQUID LINE

Liquid flashing, or turning to vapor, in the liquid line may be due to a shortage of refrigerant, a liquid line that is too small, or a liquid line that has too high a vertical lift. Liquid flashing will be indicated by a hissing or gurgling noise at the expansion valve and a lower than normal suction pressure.

2.38.1 Shortage of Refrigerant

When a system is short of refrigerant, the leak should be found and repaired and the system charged to the proper level with the correct refrigerant and checked for proper operation. Be sure to observe all safety precautions during this process.

2.38.2 Liquid Line Too Small

A liquid line that is too small will offer an excessive amount of resistance to the flow of refrigerant. As a result, the refrigerant pressure will drop near the outlet of the line, resulting in evaporation of a portion of the refrigerant and the picking up of heat outside the space being cooled, thus reducing the effectiveness of the system. There will also be a drop in temperature of the liquid line near the evaporator. This drop in temperature can be felt with the hand (see Figure 2-89).

Oil return may also be a problem when an undersized liquid line is used. When too small a liquid line is found, it must be replaced with one of proper size. Manufacturers provide chats that can be used to properly size refrigerant lines.

2.38.3 Vertical Position of Liquid Lines

When liquid lines extend vertically for more than approximately 20 ft. (6.09 m), the weight of the liquid forcing downward will probably cause flash gas. This flash gas is due to the drop in pressure. Example: if liquid Refrigerant 22 is forced up 30 ft. (9.144 m), the pressure on top of the liquid will be 15 psig (103.42 kPa) lower than that at the bottom of the column. This pressure difference plus the heat in the liquid will cause flash gas. When this condition occurs, a lower than normal suction pressure and a reduction in temperature of the liquid line near the evaporator will be present. To solve

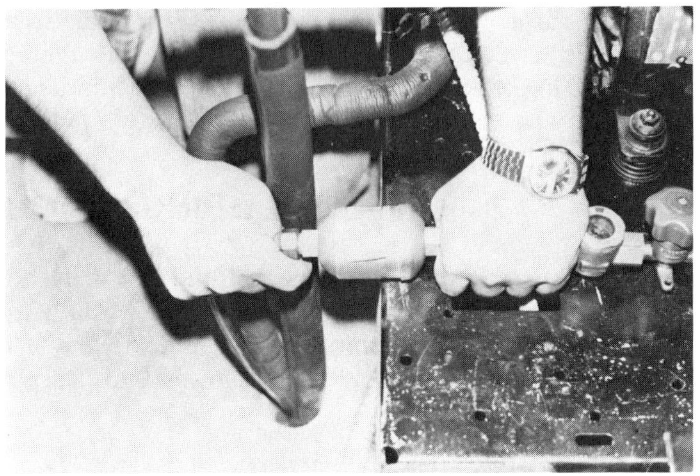

Figure 2–89 Feeling line for temperature drop.

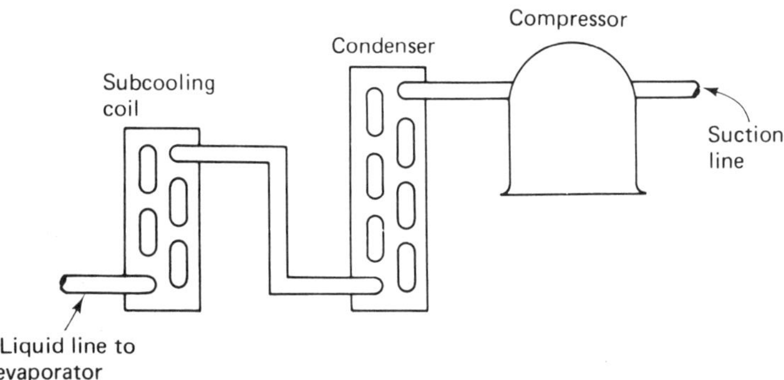

Figure 2–90 Subcooling coil location.

this problem, the liquid refrigerant must be subcooled. This may be accomplished by installing another condenser coil in the liquid line near the condenser outlet, or in severe cases, a suction to liquid heat exchanger could be added to the system (see Figure 2–90). To prevent overloading the compressor, use the proper size heat exchanger.

2.39 RESTRICTED OR UNDERSIZED REFRIGERANT LINES

Restricted or undersized refrigerant lines can be a source of continual trouble in an otherwise carefully selected unit. Often the system capacity is reduced, and oil return problems are frequently experienced with undersized lines. When these two conditions are repeated, a reevaluation of the refrigerant line sizing should be done. There are line sizing charts available for almost every situation encountered. These charts should be consulted and the lines replaced with the proper size, or an alternate recommended multiple set of lines installed.

2.39.1 Restricted Refrigerant Lines

These are usually due to contaminants or kinks in the system. A restriction in a line can be located by the difference in temperature on each side of the restriction. A restriction may be found in a drier, filter, strainer, or some other system component. This condition will be accompanied by pressure that are lower than normal and a reduction in system capacity. The restriction must be removed before proper operation can be realized. Be sure to practice safety precautions when removing restrictions from the system.

2.40 LIQUID LINE SHUT-OFF (KING) VALVES

Liquid line shut-off valves are placed between the liquid line and the outlet of the receiver tank or the condenser, depending on whether or not a receiver tank is used (see Figure 2-91).

These valves are used in service operations. Their most common use is to aid in pumping the system down. When the valve is back-seated (open), the refrigerant will flow through with very little resistance. However, when the valve is front-seated, a complete stoppage of refrigerant occurs. If the valve

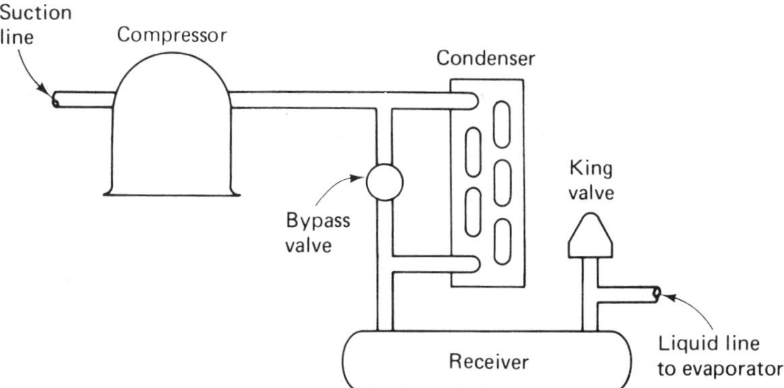

Figure 2–91 Head pressure control valve and king valve location.

is partially closed, there is a restriction to the flow of refrigerant. This restriction allows expansion of the refrigerant outside the conditioned space. This expansion may be indicated by a temperature difference across the valve, a lower than normal suction pressure, and a lower than normal discharge pressure. The obvious solution is to back-seat the valve.

2.41 COLD UNIT LOCATION

A cold unit location occurs when an air conditioning unit is operated during the winter season. This cold condition results in an excessively low discharge pressure. An excessively low discharge pressure produces two problems: (1) the pressure required to cause condensation of the refrigerant in the condenser may not be provided, and (2) the required pressure drop across the flow control device may not be provided. Both problems cause poor refrigeration. There are basically two solutions to this cold unit location: (1) provide a warmer condensing medium, and (2) install a head pressure control.

2.41.1 Operating a Small Unit in Winter

When it is necessary to operate a small air conditioning unit during the winter season, it may be possible to install the condensing unit indoors. This location will provide a warm enough condensing medium to keep the discharge pressure sufficiently high for economical operation.

2.41.2 Using Larger Units in Winter

When larger air conditioning units are used, an indoor location may be impossible. In this situation, other means of controlling the discharge pressure must be used:

1. Refrigerant pressure control valves may be used to direct the flow of hot gas around the condenser and into the receiver. During periods of low ambient air temperatures, the discharge pressure drops until the setting of the valve is reached. The flow of refrigerant from the condenser is stopped, allowing a liquid build-up in the condenser, raising the condensing pressure. After a sufficient pressure build-up, the hot gas from the compressor is discharged into the receiver, warming the liquid being discharged from the condenser. Warming the liquid will cause an increase in pressure, which will allow the flow control device to function properly. The manufacturer's recommendation for the particular type of control in question should be consulted for adjustment and service procedures.

2. The discharge pressure may also be controlled by controlling the amount of cooling medium. This may involve the restriction or complete stoppage of the air or water used to cool the condenser. When an air-cooled condenser is used, the cycling of the condenser fans in response to the discharge pressure may be used (see Figure 2-92). Or, restricting the air flow through the condenser may be used (see Figure 2-93).

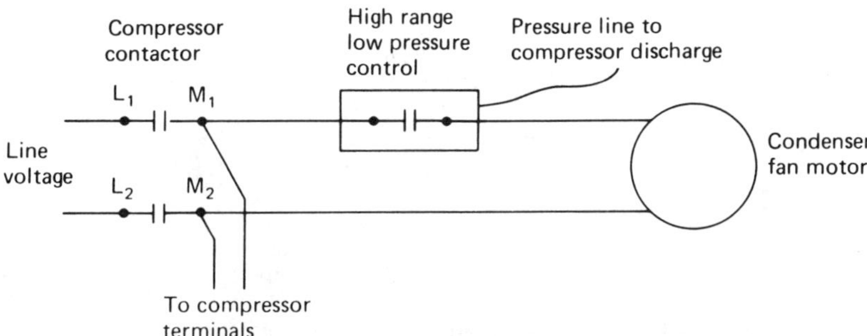

Figure 2–92 Control of head pressure by controlling condenser fan motor.

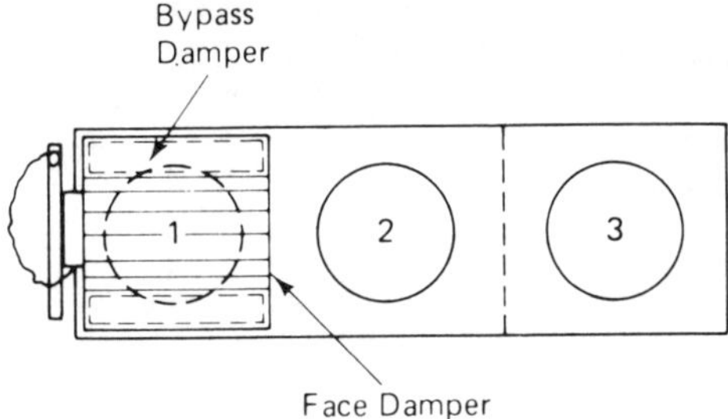

(a) Damper control location

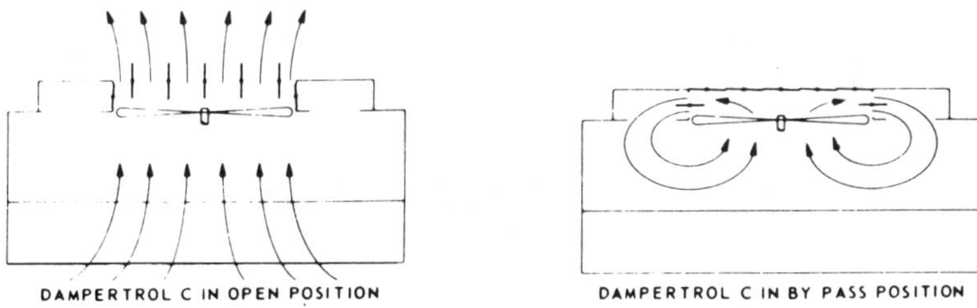

(b and c) Dampertrol C head pressure control selections: air flow in open (normal position) and closed (winter start).

Figure 2–93 Damper control operation. (Courtesy of McQuay Group, McQuay Perfex, Inc.).

The dampers are operated in response to the discharge pressure. The electrical connections may be made as shown in Figure 2-94.

The pump on a water cooled condenser may be cycled in response to the discharge pressure (see Figure 2-95). An alternate and more accurate method is to use a modulating valve and control to regulate the flow of water through the condenser (see Figure 2-96).

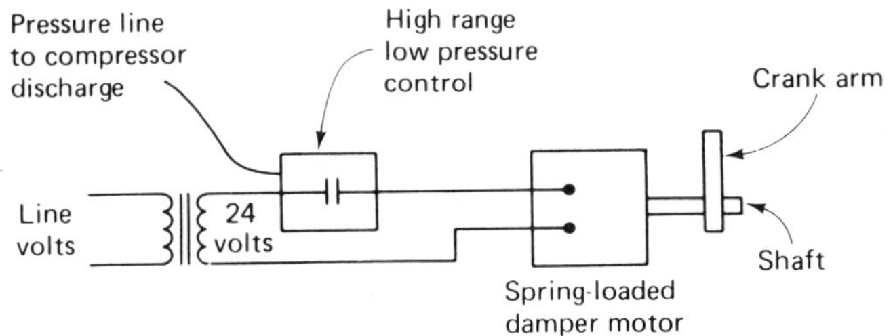

Figure 2–94 Wiring diagram for varying air head pressure control.

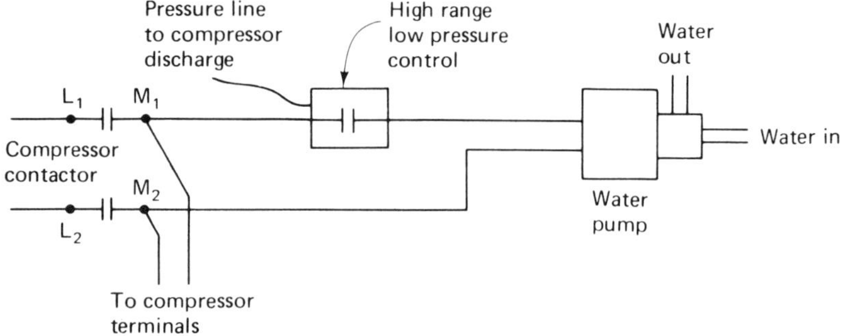

Figure 2–95 Head pressure control by cycling water pump.

2.42 EVAPORATOR TOO LARGE

The purpose of the evaporator is to absorb heat from the space being cooled and pass it on to the refrigerant. An evaporator that is improperly sized or one that absorbs too much heat due to an excessive volume of air passing through it will cause the compressor to be overloaded. An evaporator that overloads the system will cause the suction pressure to be higher than normal. The humidity inside the conditioned space will be higher than

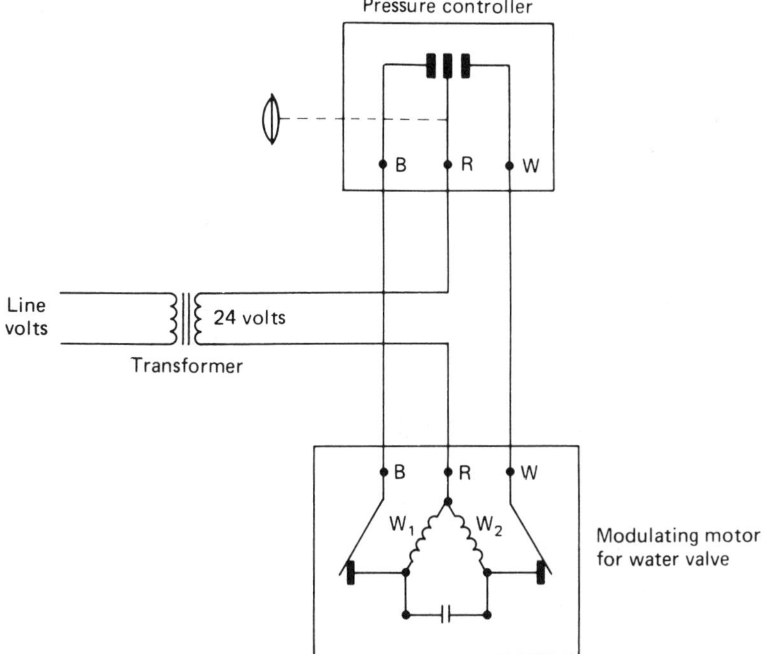

Figure 2–96 Head pressure control by use of modulating motor and water valve.

normal, and the amperage draw of the compressor will be excessive. The best approach to correcting this problem is to install a properly sized evaporator. An alternate method that may be used with less success is to reduce the air flow over the evaporator coil. The air flow should be reduced until the suction pressure is lowered to the proper pressure or until the evaporator starts to collect frost. If frost appears on the evaporator, the air flow should be increased. Use a thermometer to determine the temperature drop of the air across the coil. If the desired temperature drop is reached without frost on the evaporator, the system should operate satisfactorily.

2.43 EVAPORATOR UNDERLOADED

An underloaded evaporator does not absorb the desired amount of heat into the refrigeration system. An underloaded evaporator is generally the result of frost, a dirty coil, or a reduced air flow. When this condition occurs, a lower than normal suction pressure will be noticed. To correct this problem, the ice or dirt must be removed from the evaporator, and from the blower if it too is dirty. After all the dirt and ice have been removed, the system should be started and the temperature drop of the air crossing the evaporator checked. If the desired temperature drop and suction pressure are reached, the system will function satisfactorily. If these two readings are not normal, the situation has not been corrected and further cleaning or inspection is required.

2.44 SUCTION PRESSURE TOO HIGH

A suction pressure that is too high is generally due to excessive air flow over the evaporator, an overcharge of refrigerant, or bad suction valves in the compressor.

2.44.1 Check the Compressor Suction Valve

The first step is to check the compressor suction valves. To do this, install the compound gauge on the compressor suction service valve. (See Figure 2-97) With the unit running, front-seat the compressor suction service valve. Observe the pressure on the compound gauge. After the pressure has fallen as low as possible, stop the compressor and observe the compound gauge. The pressure should increase only slightly. If the pressure increases to zero psig (100.989 kPa), start the compressor again and reduce the pressure as much as possible. The suction pressure should be a 26 in vacuum (13.74 kPa) or lower. Stop the compressor and observe the compound gauge. The gauge should not indicate an increase in pressure of more than 5 in vacuum (17.187 kPa). If a greater increase than this is shown, replace the compressor valves, and the valve plate when replacing any one of the three. If the unit has a hermetic compressor, the complete compressor must be replaced.

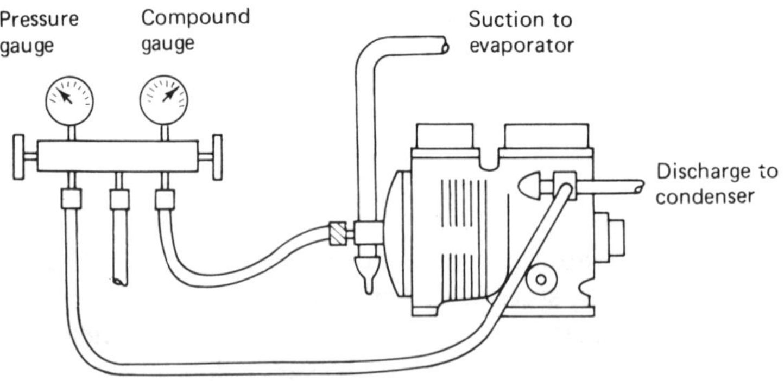

Figure 2–97 Gauges installed on a compressor.

2.44.2 An Overcharge of Refrigerant

This will cause both the suction and sicharge pressure to be higher than normal. An overcharge of refrigerant will cause approximately one-half the condenser tubes to be cool to the touch. The cool tubes are full of subcooled liquid refrigerant, which must be removed from the system. When removing the excess refrigerant, purge the refrigerant slowly and in small amounts to prevent removing too much refrigerant. When the proper charge of refrigerant is reached, only the bottom two or three rows of condenser tubing will be cool. The others will be warm or hot to the touch (see Figure 2-98).

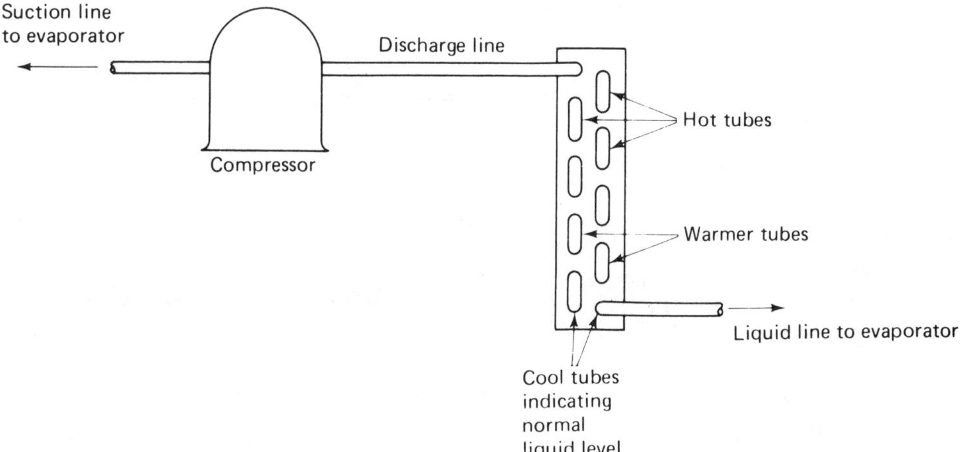

Figure 2–98 Temperature of a normally working condenser.

2.44.3 An Excessive Amount of Air Flowing over the Evaporator

This will have the same effect as too large an evaporator. The system will be overloaded by the extra amount of heat absorbed into the system. To check for this condition, measure the temperature drop across the evaporator (see Figure 2-99).

If the temperature drop is less than that desired, the air flow must be reduced to provide the proper temperature drop. This can be done by reducing the fan speed, replacing the motor with a lower rpm-type motor, or, in extreme cases, blocking the return air flow will help the situation.

2.45 MOISTURE IN SYSTEM

Moisture that has entered the system can cause serious problems. The most obvious problem is the freezing of the moisture in the flow control orifice.

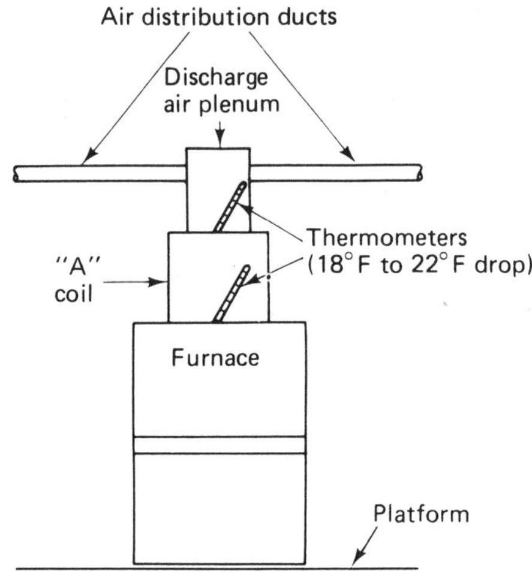

Figure 2–99 Checking temperature drop across an air conditioning evaporator.

Other more serious problems are: (1) acid forming, and (2) sludging of the compressor lubricating oil.

2.45.1 Moisture Freezing in the Refrigerant Flow Control Orifice

This will result in poor refrigeration, accompanied by a lower than normal suction pressure and possibly a lower than normal discharge pressure. To determine if moisture is freezing in the flow control device, stop the compressor. Install a compound gauge on the compressor suction service valve. Warm the flow control orifice area with a rag wet with warm or hot water while observing the compound gauge. *Caution:* Never heat a flow control device with a welding torch because of possible damage to the flow control device. If ice is causing the problem, it should melt within a few minutes, allowing the refrigerant to pass through, causing a rapid increase in the low side pressure.

There may also be a hissing noise in the flow control device while the refrigerant is flowing through. To correct this situation, install new driers in the system. It is sometimes desirable to oversize the new driers to remove the moisture. In severe cases, purging of the complete refrigerant charge, triple-evacuation of the system, and installation of driers in both the liquid and suction lines are required, along with a new charge of dry refrigerant. The system should be checked at least every 24 hours for several days and the driers replaced when necessary to completely remove all the moisture.

2.45.2 Acid Forming in a System Due to Moisture

This will cause serious damage to the compressor and expansion valve parts and will attack the insulation on the motor windings in a hermetic compressor. The preliminary indication of acid caused by moisture is copper plating (a copper color on the steel valves and parts) of the system components (see Figure 2-100).

More advanced stages are indicated by motor burnout or repeated motor burnout. A discoloration of the oil will also be noticed. This condition may be present without freezing of the refrigerant flow control device. To correct this problem, replace the refrigerant driers and check the oil color every 24 hours for several days. Replace the driers when needed until the oil color returns to normal.

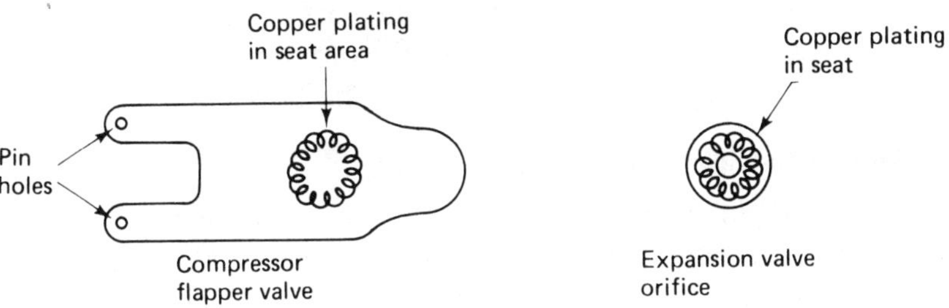

Figure 2–100 Copper plating of steel seats.

2.45.3 Sludging of the Compressor Oil

This is usually due to contaminants in the system. Oil that has started sludging will not provide proper lubrication to the system. This condition will usually be accompanied by copper plating (a copper color on the steel valves and parts) (see Figure 2-100). To correct this situation, remove all the oil from the system and charge the proper amount of fresh clean oil into the compressor. Replace all the refrigerant driers and check the system at least every 24 hours for several days. Replace the driers when needed until the color of the oil returns to normal.

2.46 ELECTRIC MOTORS

Electric motors are used for many purposes in the refrigeration and air conditioning industry other than to drive the compressor. The majority of uses for these motors include fan motors and pump motors. There are several different types of motors in use. They are: (1) split-phase (SP); (2)

permanent split capacitor (PSC); (3) capacitor start (CS); (4) capacitor start capacitor run (CSCR); and (5) shaded pole. The amount of starting and running torque required to do the job will determine what type motor is used. The steps listed below are used to check electric motors. If any of the following conditions are found, the motor must be either repaired or replaced.

2.46.1 Open Motor Windings

These occur when the path for electric current is interrupted. This interruption occurs when the insulation on the wire deteriorates and allows the wire to overheat and burn apart.

To check for an open winding, remove all external wiring from the motor terminals or connections. Using an ohmmeter, check the continuity from one terminal to another (see Figure 2-102). Be sure to zero the ohmmeter. The open winding will be indicated by an "infinity" resistance reading on the ohmmeter. There should be no continuity from any terminal to the motor case. Repair or replace the motor found to be faulty.

2.46.2 Shorted Motor Windings

These occur when the path for electric current is interrupted. This interruption occurs when the insulation on the wire deteriorates and allows the wire to overheat and burn apart. (See Figure 2-101)

In some instances, depending on how much of the winding is bypassed, the motor may continue to operate but will draw excessive amperage. To check for a shorted winding, remove all external wiring from the motor terminals. Using an ohmmeter, check the continuity from one terminal to another (see Figure 2-104).

Be sure to zero the ohmmeter. The shorted winding will be indicated by less than normal resistance. In some cases it will be necessary to consult the motor manufacturer's data for the particular motor to determine the correct resistance requirements. There should be no continuity from any terminal to the motor case. Repair or replace the motor found to be faulty.

2.46.3 Grounded Motor Windings

Grounded motor windings occur when the insulation on the winding has broken down and the winding becomes shorted to the motor housing (see Figure 2-105). Be sure to zero the ohmmeter. The grounded winding will be indicated by a resistance reading from the terminal to the case (see Figure 2-106). It may be necessary to remove some of the paint or scale from the motor case so that an accurate reading can be obtained. If a reading is obtained, repair or replace the motor.

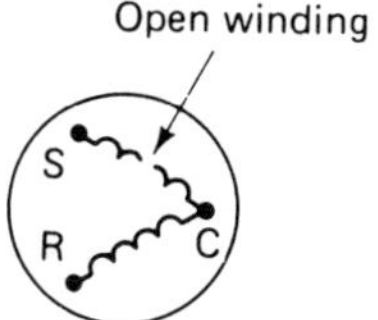

Figure 2–101 Open motor windings.

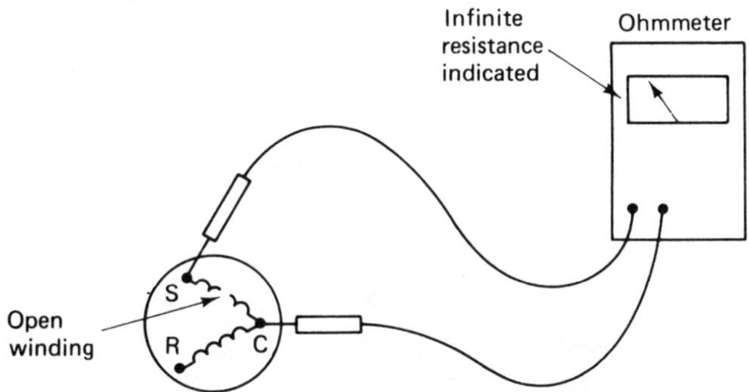

Figure 2–102 Checking continuity of open motor winding.

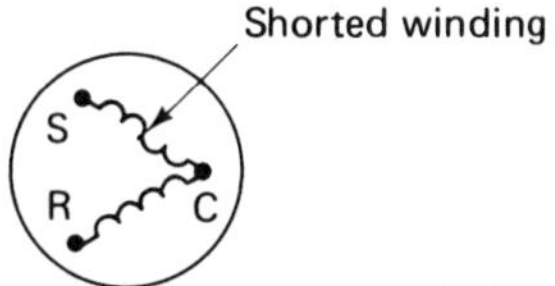

Figure 2–103 Shorted motor winding.

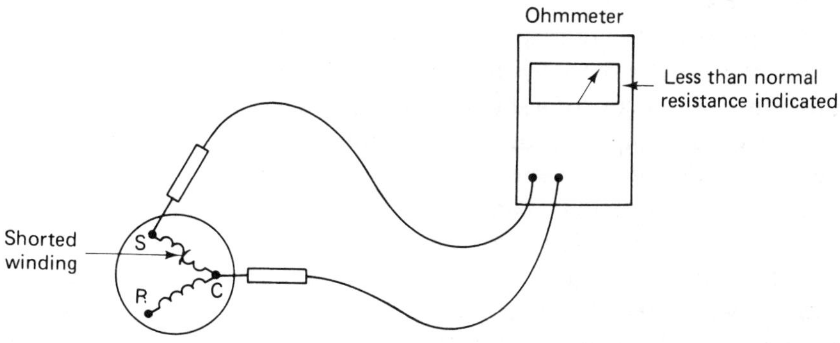

Figure 2–104 Checking continuity of shorted motor winding.

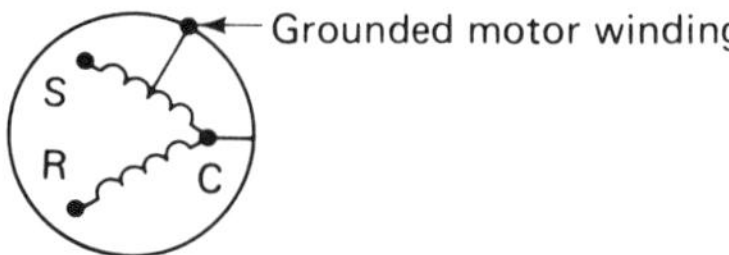

Figure 2–105 Grounded motor winding.

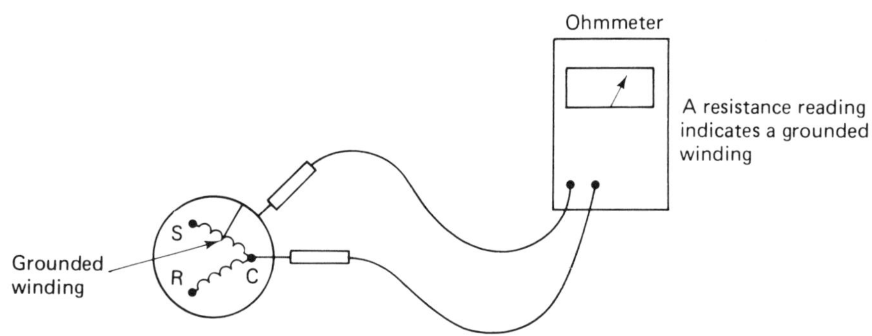

Figure 2–106 Checking for grounded motor winding.

2.46.4 Bad Starting Switch

A bad starting switch on a split-phase motor will keep it from starting properly. This switch provides a path for electrical power to flow to the starting (auxiliary) winding during the starting period, and it interrupts the electrical power when the motor has reached approximately 75 percent of its running speed. The contacts on these switches may become stuck closed, pitted, or stuck open.

A starting switch that is stuck closed will allow the motor to start but will cause the overload to open and stop the motor after a short period of operation. To check for a starting switch that is stuck closed, start the motor and check the amperage draw. This amperage draw will not drop when 75 percent of the running speed is reached. If there is any drop in amperage, check for an overload or bad bearings. If the amperage draw does not drop, have the starting switch replaced.

A starting switch that is pitted or stuck open will not allow the motor to start or will allow it to run backwards. A pitted or starting switch that is stuck open will, however, allow the motor to hum and try to start. When this condition is encountered, start the motor turning while it is humming. If the motor comes up to speed, the amperage draw is normal, and the motor operates normally, have the starting switch replaced. If the motor sometimes runs in the wrong direction, the starting switch is probably the cause. Have the starting switch replaced.

2.46.5 Bad Bearings

Bad bearings in a motor will present an overloaded condition, causing an excessive current draw. The motor may stop due to the overload after operating for a while. To check for bad bearings, remove the belt if one is used, and free the motor shaft. Move the motor shaft in a sideways movement with the hand, not end to end (see Figure 2-107). If sideways movement in the shaft is found, either have the bearings replaced or replace the entire motor.

2.46.6 Replacing the Motor

Motor replacement should be done with great care. An exact replacement must be found to be certain that the system efficiency is maintained. Be certain that the motor manufacturer's wiring diagram is followed. After tightening all the motor mounts, check to be sure that the shaft is free to rotate. Start the motor momentarily to see that it turns in the right direction and that nothing is dragging. Then start the motor and allow it to operate under normal conditions while checking the amperage draw. If the amperage draw is excessive, the reason should be found and corrected before leaving the job.

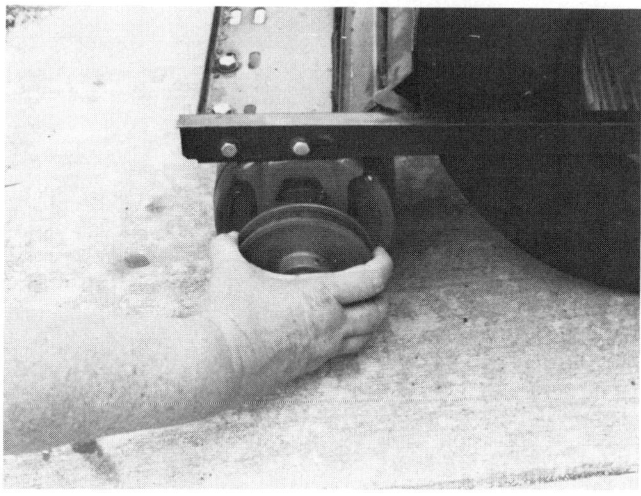

Figure 2–107 Checking motor bearings.

2.47 TRANSFORMERS

Transformers are electrical devices that are used to reduce, or to increase, electrical voltage. In refrigeration and air conditioning work, they are used to reduce line voltage to low voltage (24 volts). See Figure 2-108.

The 24 volts are used in the control circuit because it is safer, the controls are less expensive to manufacture, and the controls are more responsive to temperature change than are line voltage controls. To check a transformer, use the voltmeter and check the input voltage to the primary side of the transformer (see Figure 2-109). If voltage is found, check the output voltage from the second side of the transformer (see Figure 2-110).

Figure 2–108 Class 2 transformer.

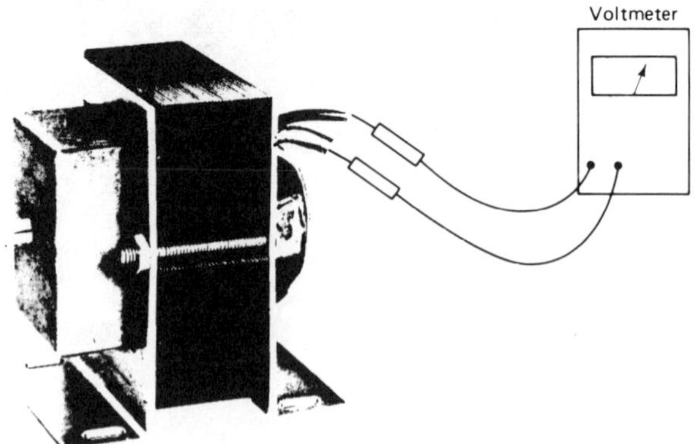

Figure 2–109 Checking voltage on a transformer primary.

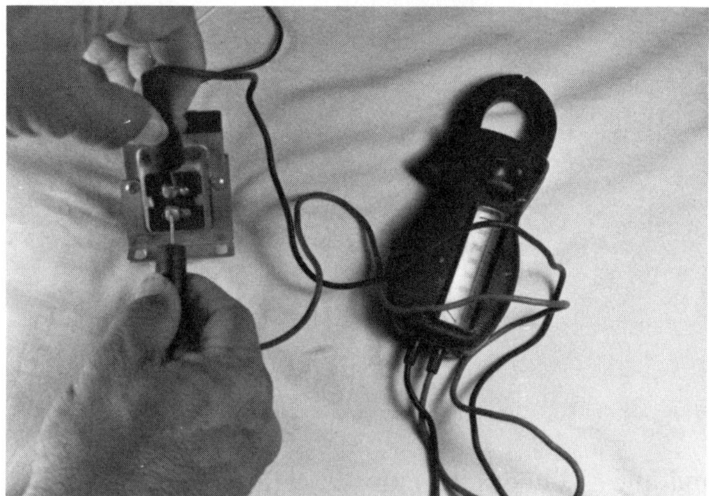

Figure 2–110 Checking the voltage on a transformer secondary.

If no secondary voltage is found and primary voltage is found, the transformer is bad and must be replaced. If no voltage is found to the primary side of the transformer, the trouble is elsewhere. An alternate method of checking a transformer is to first disconnect all external wiring to the transformer. Then check the primary and secondary wirings, in turn, with an ohmmeter. If either winding is open, the transformer is bad and must be replaced.

Some transformers have a fuse in the secondary winding. If this fuse should blow, the transformer is rendered inoperative and usually must be replaced. Fuses can be replaced in only a few transformers.

2.47.1 Replacing a Transformer

When replacing a transformer, be sure to use one with at least an equal VA (volt ampere) rating as the one being replaced. Always use a replacement designed for the same electrical characteristics as the one being replaced. Never use one with a smaller VA rating unless it is known that the replacement will have sufficient capacity. A transformer with too small a VA rating will only burn out because of an overloaded condition. Always check to be sure that there is no short to cause the replacment transformer to burn out. To check for an overload or short, measure the amperage draw in the low voltage circuit. Then multiply the amperage times the voltage to obtain the VA draw of the circuit. The circuit VA must be equal to or smaller than the transformer rating. This check must be made with the unit running. Because of the small current draw in the control circuit, it may be necessary

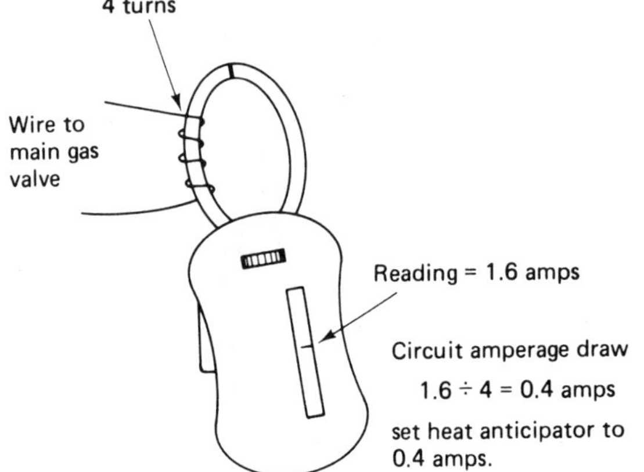

Figure 2–111 Checking amperage in a temperature control circuit.

to use a multiplier. A multiplier may be purchased or it may be handmade. To make one, simply coil a piece of wire around the tong of an ammeter (see Figure 2-111).

Connect the multiplier into the circuit and check the current draw. Then divide the current draw by the number of turns in the handmade coil. *Example:* A coil has ten turns and the current reading is 4.5 amps. Therefore 4.5 divided by 10 = 0.45 amps. The VA should be 0.45 × 24 = 10.8 VA.

2.48 HIGH AMBIENT TEMPERATURES

High ambient temperatures are generally encountered during the summertime, especially in air conditioning installations. These high temperatures result in high discharge pressures and poor operation of the unit. The best method of helping this situation is to provide a shade of some type over the unit. The shade should extend several feet past the unit toward the direction of air flow into the unit (see Figure 2-112).

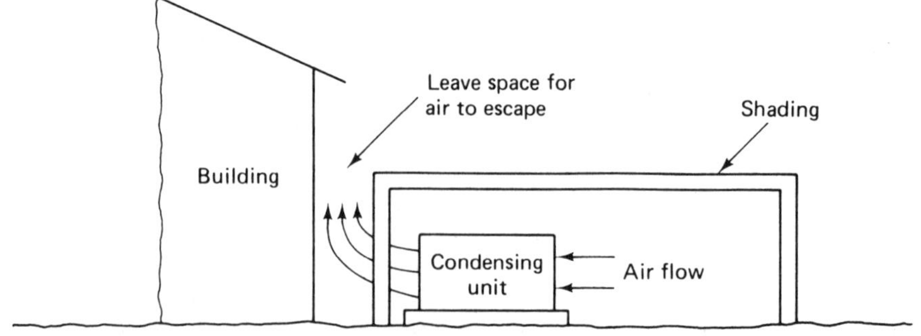

Figure 2–112 Shading for condensing unit.

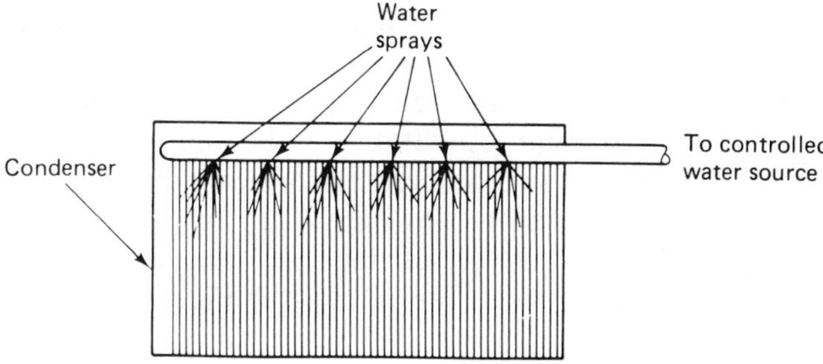

Figure 2–113 Water spray on air-cooled condenser.

An alternate method is to provide a spray of water on the condenser. This can involve an intricately designed unit or a lawn sprinkler, depending on the efficiency required (see Figure 2-113). Care should be taken to protect the electric motors and components from becoming electrically shorted to the unit.

2.49 HIGH RETURN AIR TEMPERATURE

High return air temperature is usually encountered when the unit has been shut down for a while and is then restarted after the building has reached ambient temperature. The high temperature of the air flowing over the evaporator will cause a high suction pressure, resulting in a temporary overload of the unit. The only solution to this problem is to allow the unit to operate until the return air temperature is lowered to the designed operating temperature. The unit should then operate normally.

2.50 LOW RETURN AIR TEMPERATURE

A low return air temperature usually occurs when the thermostat has been set too low. In this case the unit will run too long and cool the air below the desired temperature. The low temperature of the air flowing over the evaporator will not place a heavy enough load on the system, and the suction pressure will be lower than normal. There may be frost on the evaporator and sometimes extreme sweating of the suction line back to the compressor. The best solution to this problem is to set the thermostat to a higher setting and allow the air to warm to the designed air temperature.

2.51 INDOOR BLOWER RELAY

The purpose of an indoor blower relay is to allow the user to select continuous or intermittent operation of the indoor fan motor at the thermostat and cause the fan to operate on a slow speed during the heating season and on a higher speed during the cooling season. This type of relay is normally equipped with one set of open and one set of closed contacts. There are two general types of fan relays (see Figure 2-114).

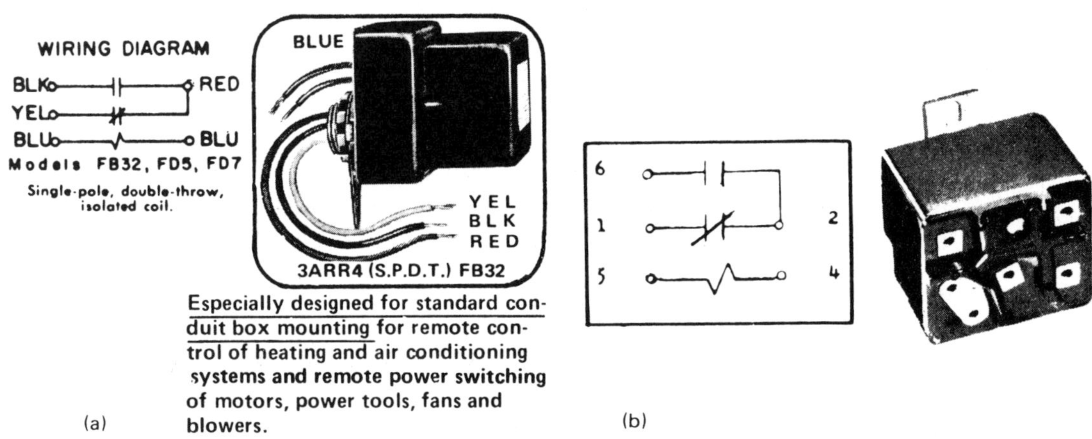

Figure 2–114 (a) Shrouded, and (b) open-type fan relays and corresponding schematics.

The open type has marked terminals, and the shrouded type has colored wires. To check out a fan relay, first turn the thermostat fan switch to "on." The relay should click. If there is no click, check the voltage on the two coil connections (see Figure 2-115).

If the voltage is indicated and no click is heard when the relay is energized, the relay is sticking and should be replaced. If a click is heard and the fan still does not run, check the line voltage across the common connection and the other two connections in turn. See Figure 2-116.

With the relay de-energized, a voltage reading should be obtained between the common and the low speed connections. If these two checks are not proven, the relay is bad and should be replaced. However, if these two checks prove the relay is good, the problem is elsewhere and should be found. When the relay must be replaced, be sure the replacement has the same voltage and amperage ratings as the one being replaced.

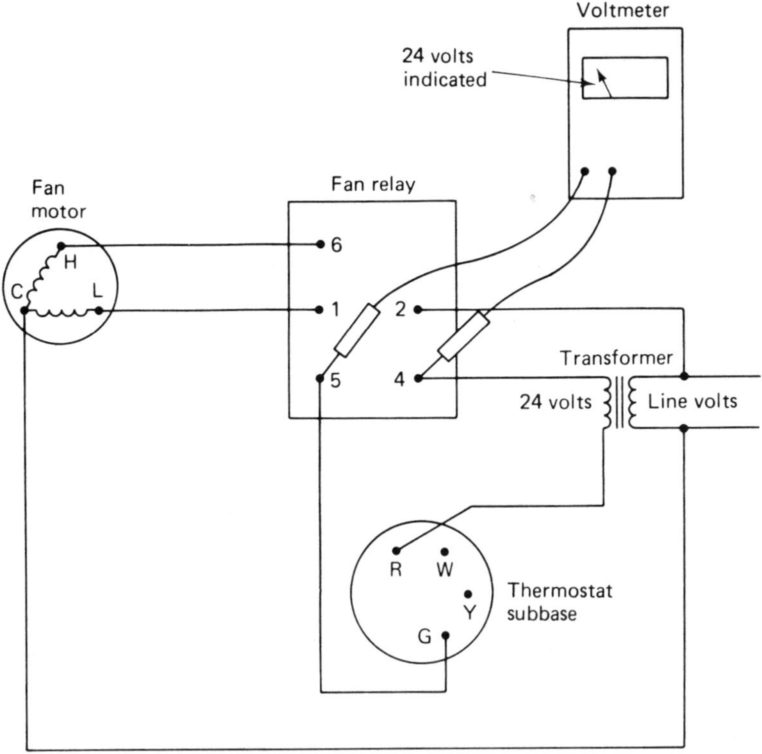

Figure 2–115 Checking voltage on fan relay coil.

2.52 AIR DUCTS

Air ducts are the hollow tubes that direct the flow of air from the air handler to the conditioned space. These ducts must be properly sized to direct the desired amount of air to the required space. Designing a practical and efficient duct system requires much time, effort, and calculation. The proper amount of insulation around the ducts is one of the important features that requires careful attention. There should be a minimum of 2 in. (25.4 mm) thickness of insulation with a vapor barrier. There are times when more insulation may be required. Enough insulation should be applied to prevent a heat loss or gain of more than about 2° F (1.11° C). See Figure 2-117.

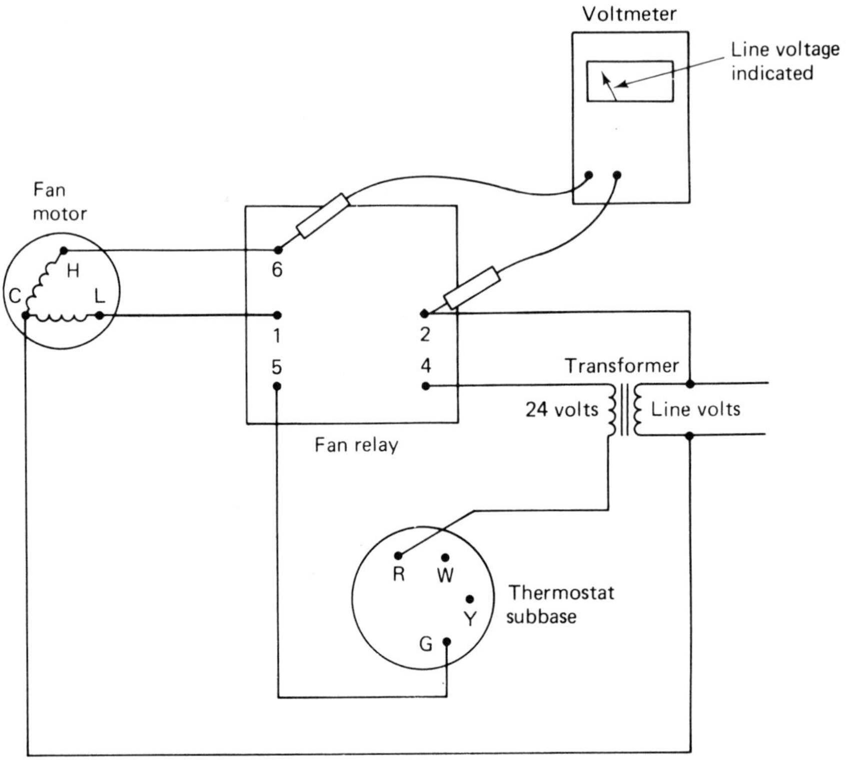

Figure 2–116 Checking voltage on fan relay contacts.

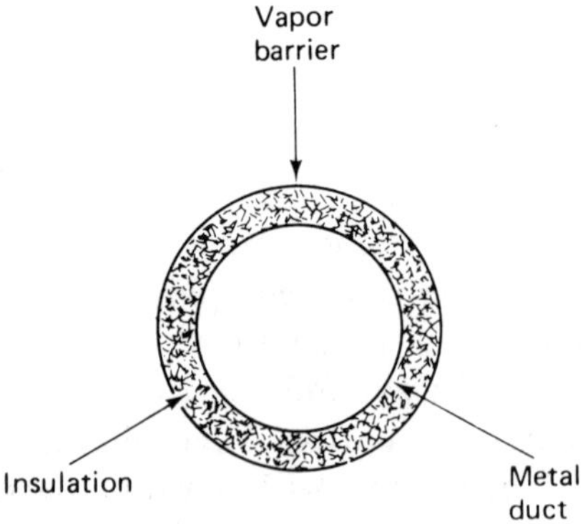

Figure 2–117 Cross section of metal duct with insulation.

2.53 UNIT TOO SMALL

Occasionally a designer will miscalculate and a wrong sized unit will be installed. When this occurs, the only remedy is to install a unit of the proper size. A unit that is too small will operate properly, but the space temperature will be higher than desired on cooling and lower than desired on heating.

2.54 OUTDOOR FAN RELAY

Outdoor fan relays are used on heat pump systems to aid in starting and stopping the outdoor fan in the various cycles required in this type of system. These relays are generally SPST (single-pole, single-throw) switches that are actuated by a 24-volt coil. The contacts are in the line voltage circuit to the fan motor (see Figure 2-118).

To check an outdoor fan relay, set the thermostat to a temperature that will demand unit operation. Cycle the unit and listen for the relay to click. No click indicates that the relay is not pulling in. Check the voltage on the relay coil connections, there should be 24-volts indicated. If not the trouble is elsewhere. If a click is heard, check the contacts. With the relay energized and the system turned on, there should not be a voltage indicated across the contacts. The relay is bad if a voltage is indicated and the fan does not run. Jumper the contacts to see if the fan will run, further proof that the relay is bad. When replacing a relay be sure that the replacement has the correct voltage and amperage rating for both the coil and the contacts.

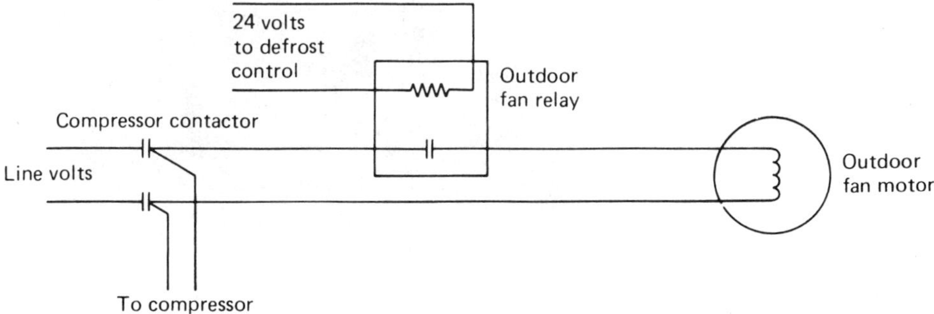

Figure 2–118 Outdoor fan relay wiring connections.

2.55 HEATING RELAY

Heating relays are used on electric furnaces and heat pump systems to complete the electrical circuit to the heating elements on demand from the thermostat or defrost control. These relays operate by an electrically operated bimetal disc switch. See Figure 2-119.

Figure 2–119 Bimetal heating relays.

The bimetal heater is in the 24-volt circuit, and the contacts are in the line voltage to the heating elements (see Figure 2-120).

To check the heating relay, use a voltmeter and check the voltage across the contacts. If a voltage is indicated here, the contacts are open. Next, jumper the contacts. If the heating element is energized, the trouble is in the heating relay. An energized heating element will be indicated by a current flow to the element. Replace the relay with an exact replacement or the unit will not operate as designed.

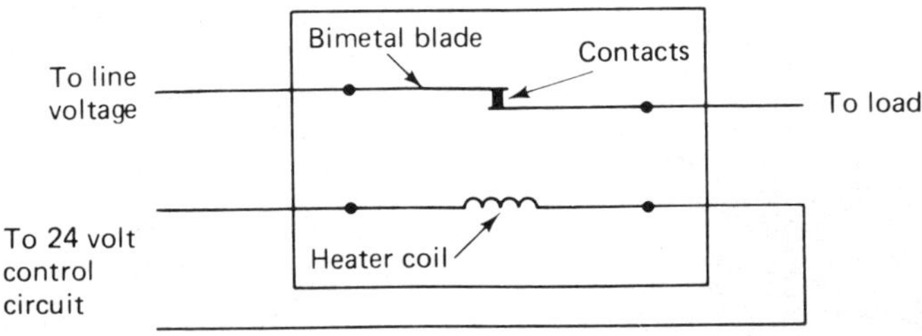

Figure 2–120 Inside of a bimetal heating relay.

2.56 BLOWERS

Blowers are the devices used to force air through the equipment and into the conditioned space (see Figure 2-121). Blowers are generally located at the air inlet to the furnace or to the air distribution system. A sufficient amount of air will not be moved if the blower vanes become clogged with dirt, if the belt becomes loose and slipping, if the bearings become worn or out of lubricant, or if the motor develops a problem.

2.56.1 Cleaning

The blower should be removed from the equipment for it to be cleaned effectively. Sometimes a brush and soapy water are required for cleaning. In such cases the blower should be taken out of the building and the motor protected from the water spray.

2.56.2 Defective Bearings

Defective bearings will overload the motor and cause the blower to run slower than is required, thus reducing the air flow and affecting the equipment operation. To check for defective bearings, remove the fan belt and grasp the blower shaft in the hand. Move the shaft from side to side, not end to end. See Figure 2-122. If sideways movement is detected, replace the

a

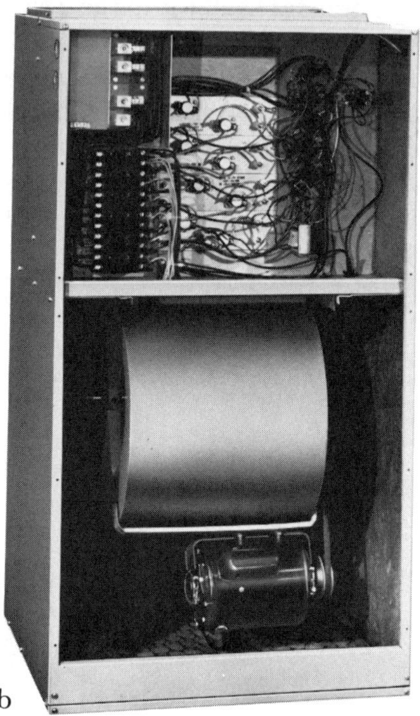
b

Figure 2–121 Furnace blower assembly (a) direct drive (b) belt drive. (Courtesy of Square D company).

Figure 2–122 Checking fan bearings.

bearings. Be sure to check the blower shaft for wear or scoring. If wear or scoring is indicated replace the shaft also. If no sideways movement is detected, spin the blower and notice if the blower turns free or if it is stiff. If a stiffness is detected, lubricate the bearings. If lubrication does not solve the problem, replace the bearings as described above. Replace the belt. Do not overtighten the belt. A belt that is too tight will cause excessive wear on the bearings. Belt adjustment is discussed in Section 2.26.2. Motor service is discussed in Section 2.46.

2.57 LOCKOUT RELAY

Lockout relays are used to interrupt the electrical circuit if an unsafe condition occurs. To reset a lockout relay, manually interrupt the electrical circuit, and then turn it back on. If the condensing unit does not come on after resetting the electrical circuit, and all other conditions are normal, check for continuity between terminals 2 and 3 with an ohmmeter. Be sure to turn off the electrical power and remove the wires from the terminals before checking the continuity. A continuity reading indicates that the contacts are stuck closed. Replace the relay with an exact replacement. See Figure 2-123.

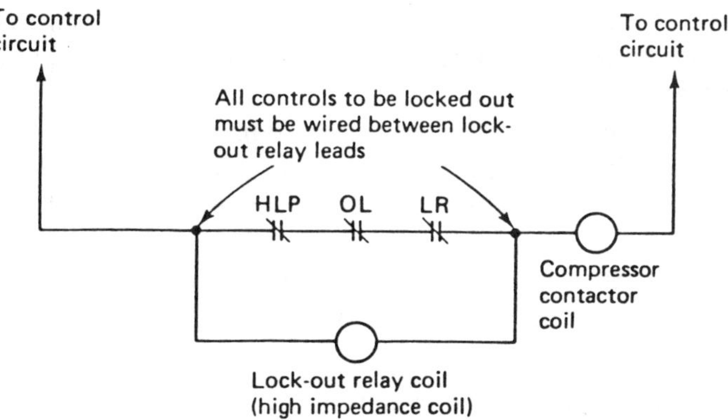

Figure 2–123 Lockout relay connections.

2.58 END PLAY IN MOTOR

Electric motors are equipped with fiber washers, or spacers, on the shaft so that alignment of the rotor and the magnetic field is possible (see Figure 2-124). These washers, or spacers, become worn and will allow the rotor to

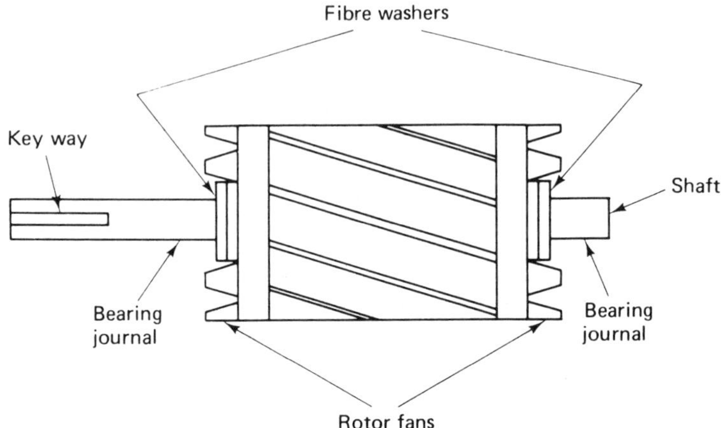

Figure 2–124 Fiber washers on motor shaft.

float back and forth in the bearings. This floating movement will cause a noise indicating that repair is needed. This repair is usually done in a motor repair shop. Therefore, the motor must be removed and delivered. When replacing the motor, follow the manufacturer's recommendations for the proper procedures.

2.59 CHECK VALVES

Check valves are used on heat pump systems to help in directing the flow of refrigerant through the proper path for the desired cycle—heating, cooling, or defrosting (see Figure 2-125).

The check valve will allow the refrigerant to pass either around the flow control device or through the flow control device. Use the following suggestions to check the operation of a check valve.

1. A check valve sticking closed in the outdoor section during the cooling cycle will cause the suction pressure to be low. The superheat will be high on the indoor coil.
2. A check valve sticking open in the indoor section during the cooling cycle will cause high suction pressure, flooding of the compressor, and low superheat on the indoor coil.
3. A check valve sticking closed in the indoor section during the heating cycle will cause the suction pressure to be low, and the superheat to be high on the outdoor coil.
4. A check valve sticking open in the outdoor section during the heating cycle will cause high suction pressure, flooding of the compressor, and low superheat on the outdoor coil.

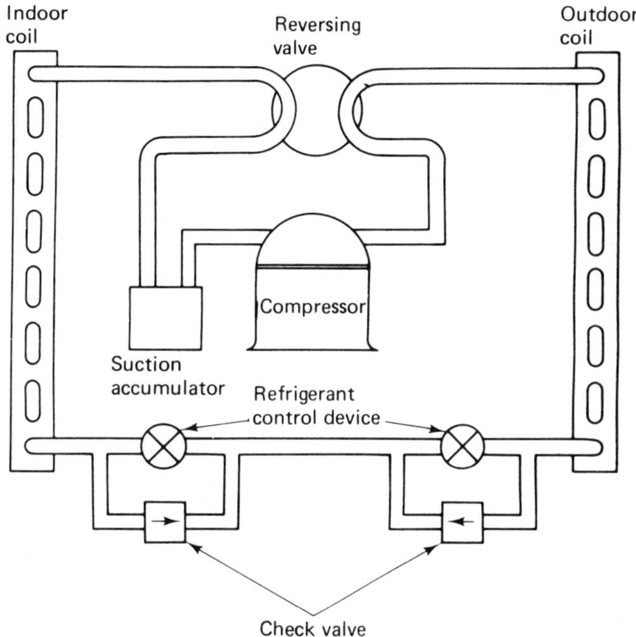

Figure 2–125 Check valve installation on a heat pump system.

If any of the above conditions are found, the check valve should be replaced. Be sure to purge the refrigerant charge or pump the system down to prevent personal injury. Make certain that the check valve is installed with the arrow pointing in the direction of refrigerant flow at that point in the circuit. See Figure 2-126.

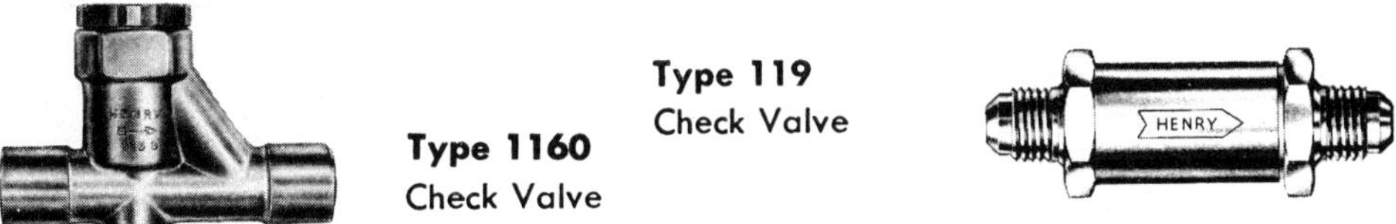

Figure 2–126 Check valves: (a) ball check, and (b) swing check. (Courtesy of Henry Valve Co.).

2.60 REVERSING VALVE

The purpose of the reversing valve is to reverse the refrigerant direction on heat pump and reverse cycle refrigeration systems (see Figure 2-127). Some systems energize the coil on heating and some energize the coil on cooling. Therefore, the system in question will determine the energized condition of

Figure 2–127 Reversing valve. (Courtesy of Ranco Controls Division).

the coil. Any troubles occurring in a heat pump system that will affect the normal operating pressures could possibly affect proper shifting of the valve. Some examples of these problems are: a refrigerant leak in the system resulting in a shortage; a compressor that is not operating at full capacity; a defective check valve; a damaged valve; or a defective electrical system. Any of these conditions will indicate an apparent malfunctioning valve. The following checks should be made on the system and its components before attempting to diagnose any valve troubles by using the "touch test" method.

1. Inspect the electrical system. This is best done by having the system in operation—either heating or cooling—so that the reversing valve solenoid is energized. While the unit is operating, remove the lock nut from the solenoid cover. Pull the solenoid coil part way off the stem. There should be a magnetic force trying to hold the coil on the stem. Be sure that electric power is applied to the solenoid during this test. If the coil is moved off the stem, the valve will return to the de-energized position. This change will be indicated by a clicking sound. A clicking sound will also be heard when the coil is replaced on the stem.
2. Inspect the reversing valve for physical damage. Check the solenoid coil for deep scratches, dents, and cracks.
3. Check the operation of the system against the equipment manufacturer's recommendations. Make any corrections needed.

If all of these prove that the system is operating properly, perform the "touch test" on the reversing valve, using Table 2-2. The "touch test" is performed simply by feeling the temperature difference of the six tubes on the valve and comparing these differences. Refer to the table for possible causes and corrections.

Table 2–2 Touch Test Chart (Table 2–2 TS, Courtesy of Ranco Controls Division)

Valve Operating Condition	DISCHARGE TUBE from Compressor 1	SUCTION TUBE to Compressor 2	Tube to INSIDE COIL 3	Tube to OUTSIDE COIL 4	LEFT Pilot Back Capillary Tube 5	RIGHT Pilot Front Capillary Tube 6	Possible Causes	Corrections
				NORMAL OPERATION OF VALVE				
Normal Cooling	Hot	Cool	Cool, as (2)	Hot, as (1)	*TVB	*TVB		
Normal Heating	Hot	Cool	Hot, as (1)	Cool, as (2)	*TVB	*TVB		
				MALFUNCTION OF VALVE				
Valve will not shift from cool to heat	Check electrical circuit and coil						No voltage to coil. Defective coil.	Repair electrical circuit. Replace coil.
	Check refrigeration charge						Low charge Pressure differential too high.	Repair leak, recharge system. Recheck system.
	Hot	Cool	Cool, as (2)	Hot, as (1)	*TVB	Hot	Pilot valve okay. Dirt in one bleeder hole.	Deenergize solenoid, raise head pressure, reenergize solenoid to break dirt loose. If unsuccessful, remove valve, wash out. Check on air before installing. If no movement, replace valve, add strainer to discharge tube, mount valve horizontally.
							Piston cup leak	Stop unit. After pressures equalize, restart with solenoid energized. If valve shifts, reattempt with compressor running. If still no shift, replace valve.
	Hot	Cool	Cool, as (2)	Hot, as (1)	*TVB	*TVB	Clogged pilot tubes.	Raise head pressure, operate solenoid to free. If still no shift, replace valve.
	Hot	Cool	Cool, as (2)	Hot, as (1)	Hot	Hot	Both ports of pilot open. (back seat port did not close.)	Raise head pressure, operate solenoid to free partially clogged port. If still no shift, replace valve.
	Warm	Cool	Cool, as (2)	Warm as (1)	*TVB	Warm	Defective Compressor.	
Start to shift but does not complete reversal	Hot	Warm	Warm	Hot	*TVB	Hot	Not enough pressure differential at start of stroke or not enough flow to maintain pressure differential.	Check unit for correct operating pressures and charge. Raise head pressure. If no shift, use valve with smaller ports.
							Body damage.	Replace valve.
	Hot	Warm	Warm	Hot	Hot	Hot	Both ports of Pilot open.	Raise head pressure, operate solenoid. If no shift, replace valve.

NOTES:
*Temperature of Valve Body
**Warmer than Valve Body.

Table 2–2 Cont.

Valve Operating Condition	DISCHARGE TUBE from Compressor	SUCTION TUBE to Compressor	Tube to INSIDE COIL	Tube to OUTSIDE COIL	LEFT Pilot Back Capillary Tube	RIGHT Pilot Front Capillary Tube		
	1	2	3	4	5	6	Possible Causes	Corrections
Start to shift but does not complete reversal	Hot	Hot	Hot	Hot	*TVB	Hot	Body damage.	Replace valve.
							Valve hung up at mid-stroke. Pumping volume of compressor not sufficient to maintain reversal.	Raise head pressure, operate solenoid. If no shift, use valve with smaller ports.
	Hot	Hot	Hot	Hot	Hot	Hot	Both ports of Pilot open.	Raise head pressure, operate solenoid. If no shift, replace valve.
Apparent leak in heating	Hot	Cool	Hot, as (1)	Cool, as (2)	*TVB	**WVB	Piston needle on end of slide leaking.	Operate valve several times then recheck. If excessive leak, replace valve.
	Hot	Cool	Hot, as (1)	Cool, as (2)	**WVB	**WVB	Pilot needle and piston needle leaking.	Operate valve several times then recheck. If excessive leak, replace valve.
Will not shift from heat to cool	Hot	Cool	Hot, as (1)	Cool, as (2)	*TVB	*TVB	Pressure differential too high.	Stop unit. Will reverse during equalization period. Recheck system.
							Clogged Pilot tube.	Raise head pressure, operate solenoid to free dirt. If still no shift, replace valve.
	Hot	Cool	Hot, as (1)	Cool, as (2)	Hot	*TVB	Dirt in bleeder hole.	Raise head pressure, operate solenoid. Remove valve and wash out. Check on air before reinstalling, if no movement, replace valve. Add strainer to discharge tube. Mount valve horizontally.
	Hot	Cool	Hot, as (1)	Cool, as (2)	Hot	*TVB	Piston cup leak.	Stop unit, after pressures equalize, restart with solenoid deenergized. If valve shifts, reattempt with compressor running. If it still will not reverse while running, replace valve.
	Hot	Cool	Hot, as (1)	Cool, as (2)	Hot	Hot	Defective Pilot.	Replace valve.
	Warm	Cool	Warm, as (1)	Cool, as (2)	Warm	*TVB	Defector compressor.	

NOTES:
*Temperature of Valve Body.
**Warmer than Valve Body.
VALVE OPERATED SATISFACTORILY PRIOR TO COMPRESSOR MOTOR BURN OUT—caused by dirt and small greasy particles inside the valve. To CORRECT: Remove valve, thoroughly wash it out. Check on air before reinstalling, or replace valve. Add strainer and filter-dryer to discharge tube between valve and compressor.

When replacing the reversing valve, the refrigerant will need to be purged. Follow the manufacturer's recommendations when installing a new valve. Evacuate the system, install new filter-driers, and recharge the system according to the manufacturer's recommendations.

2.61 DEFROST CONTROLS

Heat pump defrost controls are used to detect ice and frost on the outdoor coil during the heating cycle. The most popular method of defrost control is the time temperature method (see Figure 2-128). When using this method both the time and the temperature sections must demand defrost before the defrost cycle can be initiated. When the defrost cycle is initiated, the reversing valve changes and the outdoor fan motor stops running. The clock motor operates only when the compressor is running. The timer can be set to initiate the defrost cycle on 30-, 60-, and 90-minute intervals. If it is found that the 90-minute cycle allows too much frost on the outdoor coil, then the timer can be adjusted to provide the proper amount of defrost time. The defrost cycle can be initiated only by demand from both the time and the temperature sections of the defrost control. The defrost control temperature will initiate the defrost cycle at temperature of approximately 30° F (-1.1.°C) and will terminate the defrost cycle at approximately 60° F (15.6° C).

The time motor is operated by 230 volts and operates only when the compressor is operating. The temperature portion of the control is attached to a return bend on the outdoor coil (see Figure 2-129). The bulb is insulated to ensure that the ambient air temperature does not affect the bulb. To check

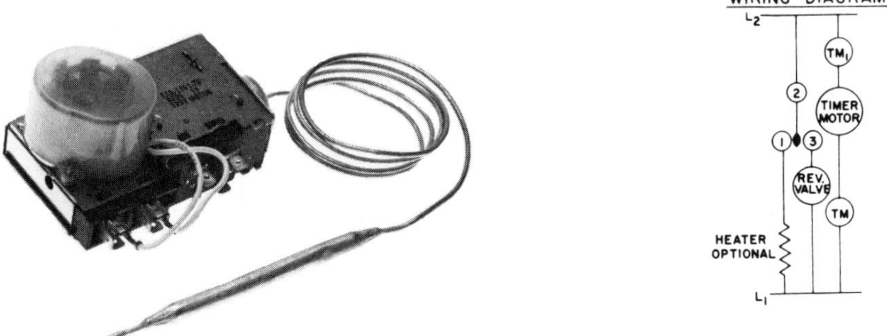

Figure 2–128 Time-temperature defrost control and schematic. (Courtesy of Ranco Controls Division).

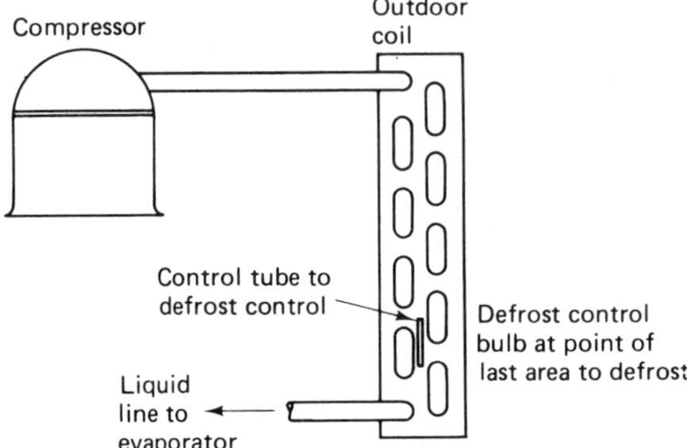

Figure 2–129 Installation of temperature bulb for defrost control.

a time-temperature defrost control, be sure that the outdoor coil is cold enough to need defrosting; then use the following steps:

1. Place a jumper between terminals 2 and 3 on the clock timer (see Figure 2-130). If the reversing valve changes, the clock is not operating. Check to be certain that power is supplied to the clock motor terminals. If power is found, the motor is defective and must be replaced.
2. If the reversing valve does not change, jumper the terminals of the defrost thermostat while the jumper remains on terminals 2 and 3. If

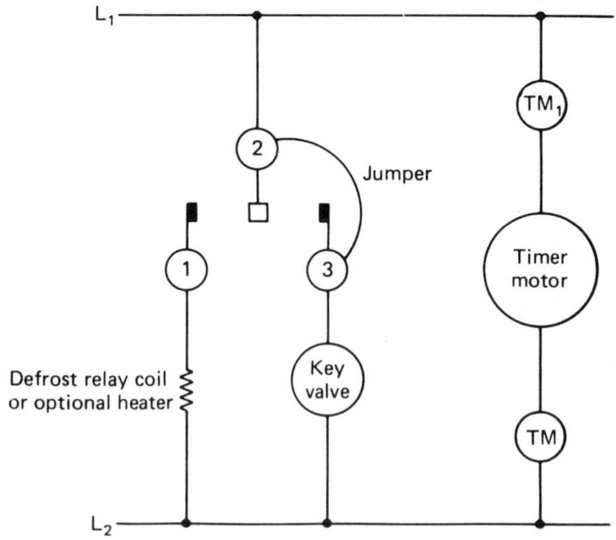

Figure 2–130 Jumper between terminals 2 and 3 on defrost timer.

the reversing valve changes, the defrost control is defective and must be replaced.

If the additional jumper does not cause the reversing valve to change, the trouble is not in the defrost control. The reversing valve solenoid coil, the reversing valve, and the defrost relay must be checked to find the trouble. Be sure that the specified voltage is applied to these controls.

2.62 AIR BYPASSING COIL

Air bypassing a coil will reduce the efficiency of the unit (see Figure 2-131). The greater the amount of air bypassing the coil, the greater the reduction in efficiency. The opening through which the air bypasses must be closed so that peak efficiency can be obtained. Air bypassing an evaporating coil causes a low suction pressure. Air bypassing a condensing coil causes a high discharge pressure.

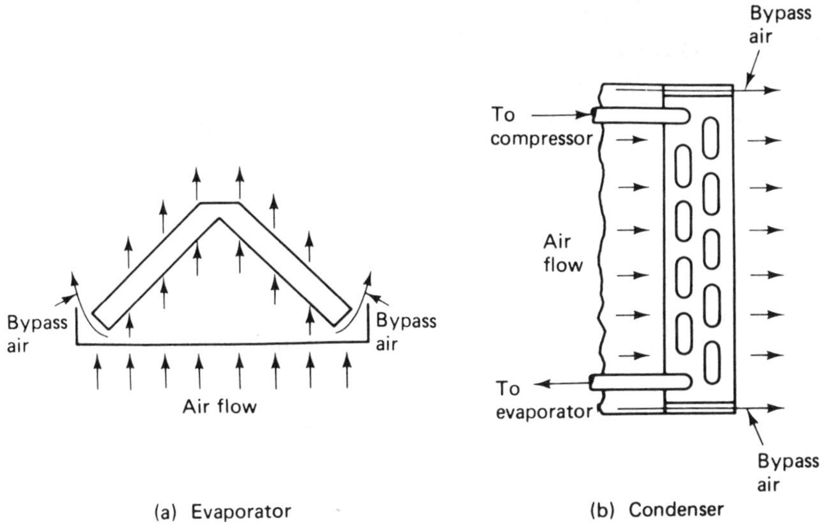

Figure 2–131 Air bypassing coils.

2.63 AUXILARY HEAT STRIP LOCATION

The auxilary heat strips must be installed downstream of the indoor coil on heat pump systems (see Figure 2-132). If the heat strips are installed upstream of the indoor coil, the heat pump will act as a reheat system. This

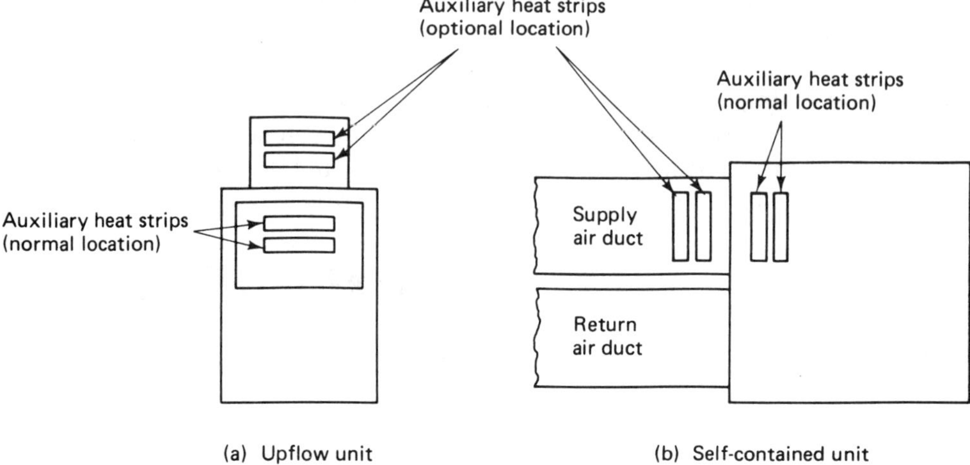

Figure 2–132 Location of auxiliary heat strips on heat pump systems.

hot air entering the indoor coil during the heating cycle will cause the discharge pressure and temperature to be excessively high, thus causing a high compression ratio on the compressor, one that will probably cause permanent damage and will require compressor replacement.

2.64 DEFROST RELAY

The purpose of the defrost relay is to cause the reversing valve to change to the cooling position, stop the outdoor fan motor, and energize the auxiliary heaters during the defrost cycle. The coil voltage is usually 230 volts. See Figure 2-133.

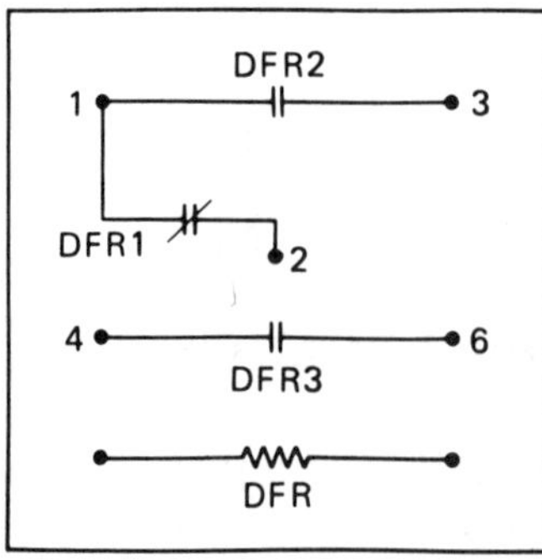

Figure 2–133 Internal schematic of a defrost relay.

This relay is energized by the defrost control. Because some manufacturers prefer to energize the reversing valve during the heating cycle and some during the cooling cycle, the wiring diagram for the particular unit in question should be used.

If the relay is energized and a click is heard when the defrost control demands defrost but the system does not shift, check the voltage across each set of contacts with a voltmeter. Open contacts will show full voltage across them. Closed contacts will show no line voltage across them. If conditions other than these are found, replace the relay.

CHAPTER 3

Start-up Procedures: Air Conditioning—Heat Pump

1. Check to be sure that electricity has been on to the outdoor unit for 24 hours.
2. Check the outdoor coil and clean if necessary.
3. Check all bearings; make necessary repairs.
4. Lubricate all bearings requiring service.
5. Check the condition and tension of all belts. Replace or adjust as required. See Figure 3-1.

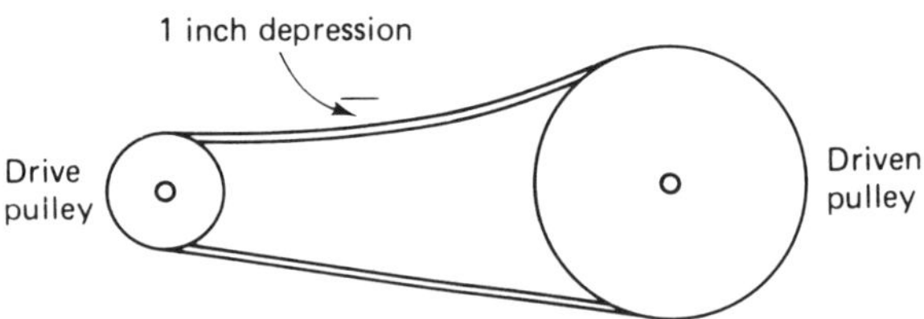

Figure 3–1 Checking belt tension.

6. Check the condition of the contactor contacts. Replace if necessary.
7. Check the tightness of all electrical connections. Tighten loose connections.
8. Check and repair burned or frayed wiring.
9. Set the thermostat for cooling, and lower the temperature setting below room temperature.

10. Check the temperature rise on the outdoor coil. See Figure 3-2.

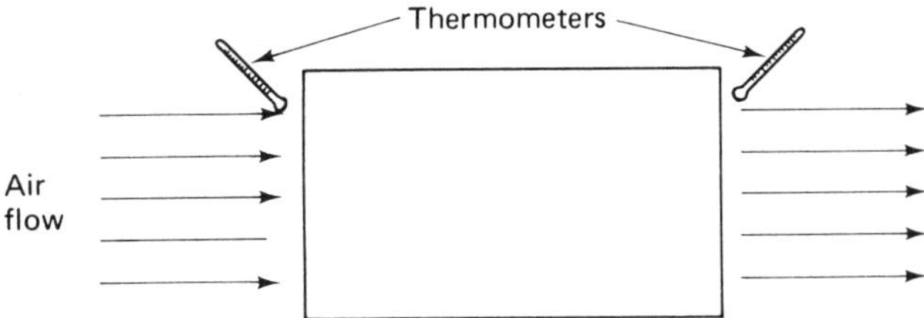

Figure 3–2 Checking temperature rise on outdoor coil.

11. Check the suction and discharge pressures.
12. Check the refrigerant charge in the system. If short, repair leak and add refrigerant.
13. Check the amperage draw to all motors.
14. Check the temperature drop on the indoor coil. If more than 24° F (13.33° C), check for a dirty coil. If tless than 18° F (10° C), check for inefficient compressor or refrigerant shortage on cooling. On heating, the temperature rise should be approximately 30° F (16.67° C). See Figure 3-3.

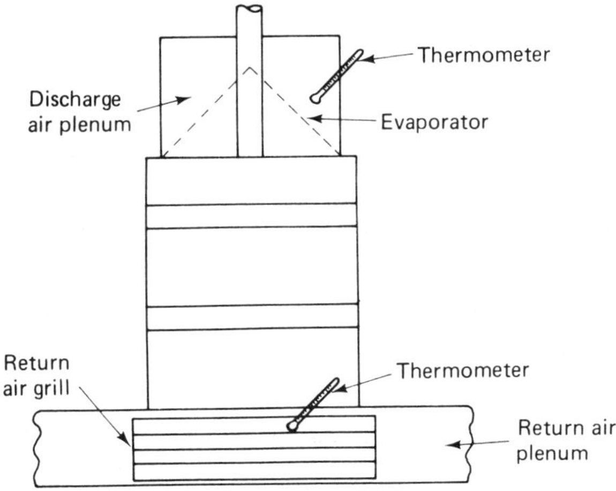

Figure 3–3 Checking temperature drop on indoor coil.

15. Check the thermostat calibration. Calibrate if necessary.
16. Clean or replace the air filter.
17. Clean and vacuum the indoor unit and the area around the unit.
18. Replace all covers.

CHAPTER 4

Standard Service Procedures

4.1 PROCEDURE FOR REPLACING COMPRESSORS

Past experiences have demonstrated that after a hermetic motor burnout has occurred, the refrigeration system must be cleaned to remove all contaminants. Without removal of these contaminants, a repeat burnout will occur. Failure to follow minimum cleaning recommendations as quickly as possible will result in an excessive risk of a repeat burnout.

Cleaning of the system by flushing with refrigerant—R-11, or similar refrigerants—has been used in the past under certain conditions and with mixed degrees of success. This flushing method is seldom used today and is not recommended here except on recommendation of the equipment manufacturer.

The first step is to be certain that the compressor is burned. The following is the general procedure used:

1. When a compressor fails to start, it may appear to be burned out. However, the fault may be in many other areas.
2. All other possibilities should be eliminated. Check for electrical and mechanical misapplication or malfunction.
3. Check the compressor motor for shorted, grounded, and open windings. If none are found, check the winding resistance with a precision ohmmeter to determine if turn-to-turn shorts exist.
4. Next, purge a small amount of refrigerant gas from the compressor discharge. Smell the gas in very small whiffs. A burned motor is usually indicated when a characteristic acrid odor is detected. Smelling the gas should be used only as a last resort.

If the motor is found burned, the refrigerant should be discharged from the system. The service technician should follow certain safety practices. In addition to the electrical hazards, the service technician should be aware of the dangers of receiving acid burns.

1. When discharging refrigerant (gas or liquid) from the system in which a burnout occurred, avoid injury by not allowing the refrigerant to touch the eyes or the skin. If the complete charge is to be purged, it should be discharged outside the building. The burned refrigerant vapors will tarnish metal surfaces as well as contaminate the air used for breathing when released inside a building.
2. When it is necessary for the service technician to come in contact with the oil or sludge in a burned-out compressor, rubber gloves should be worn to prevent possible acid burns.

When replacing the compressor in a system that has had a burnout, the system must be thoroughly cleaned to remove as many of the contaminants as possible. The following is a list of the recommended procedures:

1. On systems of 5 hp and smaller, discharge the complete refrigerant charge to the outside atmosphere—preferably in the liquid form. On units larger than 5 hp, an attempt to save the refrigerant should be made by valving off the compressor (front-seating the service valves), then purging the refrigerant from the compressor.
2. Purge the system by blowing clean dry refrigerant through the system, forcing the refrigerant completely through all piping and coils.
3. Replace the compressor. Then install an approved cleanup filter-drier in the suction line at the compressor. On units larger than 5 hp, the filter-driers cannot be installed until the refrigerant has been pumped from the low side of the system. See Figure 4-1.

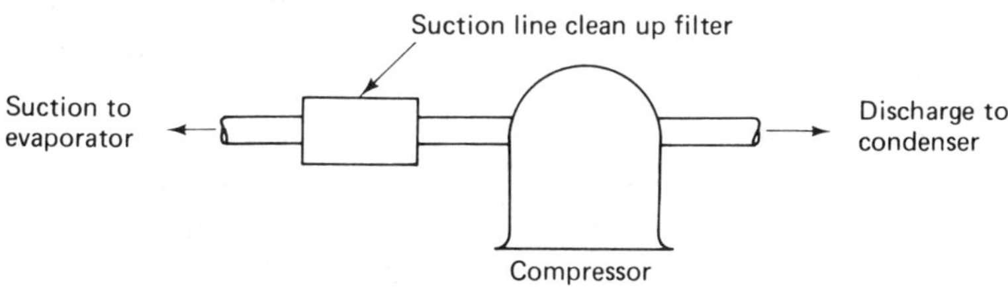

Figure 4–1 Suction line clean-up filter-drier location.

4. Install an approved oversize liquid line cleanup drier in the liquid line. See Figure 4-2. On units larger than 5 hp, this cannot be done until the refrigerant has been pumped from the high side of the system.
5. Check the refrigerant flow control device and clean it thoroughly or replace it.

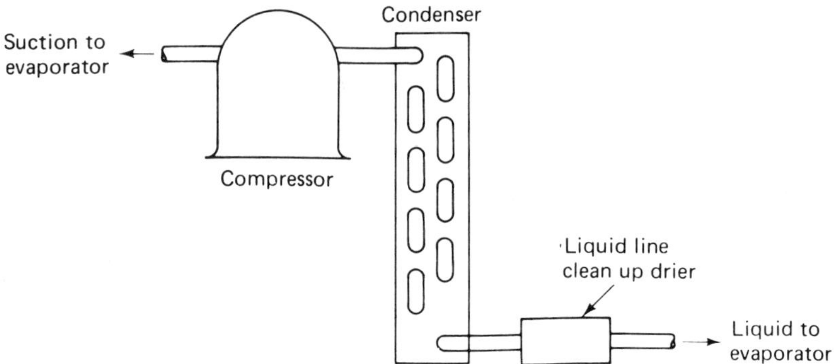

Figure 4–2 Liquid line clean-up drier location.

6. Triple-evacuate the system. The last evacuation should lower the system pressure to 1,000 microns or lower, if time permits.
7. Recharge the system and put it back into operation. On systems larger than 5 hp, add refrigerant to complete the charge.
8. After the system has been in operation for approximately 24 hours, the acid content of the oil and the condition of the cleanup driers and filters should be checked. If acid is found in the oil, noted by discoloration, or the cleanup filters and driers show signs of stoppage, they should be replaced, the oil drained, if possible, and a new charge put into the system. This procedure should be repeated until a clean system is indicated.

4.2 PROCEDURE FOR USING THE GAUGE MANIFOLD

The gauge manifold is probably the most important tool in the service technician's toolbox. The various "gauges" can be used to check the system pressure, to charge refrigerant into the system, to evacuate the system, to add oil to the system, to purge noncondensables from the system, and for many other uses.

The gauge manifold set consists of a compound gauge, a pressure gauge, and a manifold that is equipped with hand valves to isolate the different connections or to allow their use in any combination as required. See Figure 4-3. The ports to the gauge and the line connections are connected so that the gauges will indicate the pressure when connected to a pressure source. Flexible, leak-proof hoses are used to make the connections from the gauge manifold to the system.

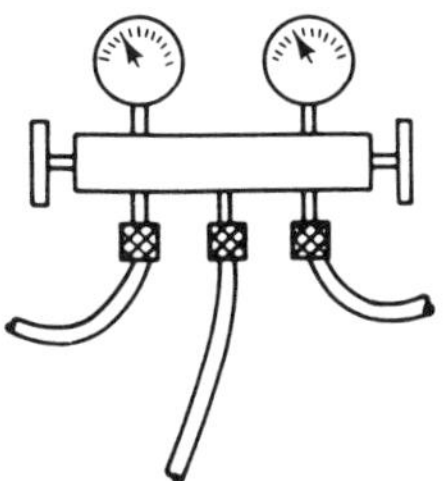

Figure 4–3 Gauge manifold.

One of the most common service functions is hooking the gauges to the system. Care should be taken to prevent contaminants from entering the system. The hoses should be purged by allowing a small amount of refrigerant to escape from the fittings before tightening. The specific procedures for connecting the gauges to a system containing refrigerant are described below.

On systems where it is certain that both pressures are above 0 psig (100.989 kPa), use these procedures:

1. Front-seat the hand valves on the gauge manifold.
2. Back-seat the system service valves. This is to isolate the gauge ports from the rest of the system.
3. Make the hose connections to the system. Loosen one end of the center hose.
4. Crack the system service valves off the back seat. Do this slowly to prevent a sudden in-rush of high pressure gas to the gauge.
5. Crack open one valve on the gauge manifold and allow refrigerant vapor to slowly escape out the center hose for a few seconds. Close the gauge manifold valve and repeat this process with the other valve.
6. The gauge manifold is now connected to the system and is ready for use. See Figure 4-4.

On systems where the low-side pressure is below 0 psig (100.989 kPa), use these procedures:

1. Front-seat the hand valves on the gauge manifold.
2. Back-seat the system service valves. This is to isolate the gauge ports from the rest of the system.
3. Make the hose connections to the system. Tighten the hose connection to the discharge service valve, loosen the hose connection to the suction service valve, and plug the center hose connection to prevent the escape of refrigerant at this point.

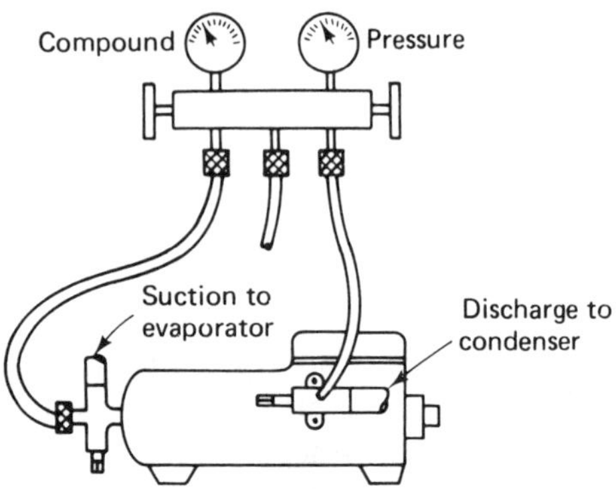

Figure 4–4 Gauge manifold connected to system.

4. Crack the system discharge service valve off the back seat. Do this slowly to prevent a sudden in-rush of high pressure refrigerant gas to the gauge.
5. Crack open the high side gauge manifold valve and allow a small amount of refrigerant vapor to escape out the low side hose connection at the system service valve. Tighten the hose connection after a few seconds. Close the gauge manifold hand valve.
6. Crack the system suction service valve off the back seat.
7. The gauge manifold is now connected to the system and is ready for use (see Figure 4-4).

Gauges are delicate instruments and should be treated with care. Do not drop the gauges or subject them to pressures higher than the maximum pressure shown on the scale. Gauges should be kept in adjustment so that accurage pressures are indicated.

4.3 PROCEDURE FOR PURGING NONCONDENSABLES FROM THE SYSTEM

Air is considered a noncondensable under the pressures and temperatures normally encountered in an air conditioning or refrigeration system. Air can enter the refrigerant circuit in several ways. The most common are a leak in the low side of the system or improper connection and use of the service gauges. In some cases it may not be practical or desirable to purge the complete refrigerant charge and evacuate the system. However, the air must be removed to prevent damage to the system due to chemical reactions and to help keep the system operating efficiently.

Normally the air will be trapped at the top of the receiver and condenser because of the liquid seal at the outlet of these components. Air in the system may be detected by a higher than normal condensing pressure caused by the trapped air. The amount of increase in pressure will be determined by the amount of trapped air.

To purge noncondensables from the system, use the following procedure:

1. Locate and remove the source of noncondensables.
2. Connect the gauges to the system (see Figure 4-4).
3. If possible, pump the system down.
4. Stop the unit. Leave the condenser fan running on air-cooled units or block the water valve open on water-cooled units and leave the pump running. Allow the unit to cool down for approximately ten minutes. During this time the noncondensables will rise to the top of the condenser.
5. If purge valves are on the unit, use them for the purging process. If not, the gauge port on the compressor discharge may be used. To purge, slowly open the purge valve. Allow the vapor to bleed off very slowly for only short periods of time to prevent boiling off of excess refrigerant, the remixing of the air and refrigerant, and the purging of an excess of refrigerant. After the system has sat idle for a few minutes, repeat the purging process. Repeat this process three or four times.
6. Start the system and check the discharge pressure after a few minutes of operation. If the discharge pressure is still abnormally high, repeat the purging process starting with step 4. Repeat steps 4, 5, and 6 until satisfactory operation is obtained.
7. Put the system back in the normal operating condition.

4.4 PROCEDURE FOR PUMPING A SYSTEM DOWN

Pumping a system down is the process generally used to put all the refrigerant into the condenser or receiver. Pumping down is used to save the refrigerant when work is required on components in the low side of the refrigerant system. Pumping a system down can be accomplished only on systems equipped with service valves and a liquid line shut-off valve.

To pump a system down, use the following procedure:

1. Install the gauge manifold on the system (see Figure 4-4).
2. Start the unit.
3. Front-seat the liquid line (king) valve.
4. Observe both the suction and discharge pressures. If there is a sharp increase in the discharge pressure, stop the compressor. Check to determine the reason for the increase. If the receiver and condenser are full of refrigerant, the remaining part of the charge must be removed from the system.

   When the suction pressure is reduced to about 1 or 2 psig (108.17 or 114.73 kPa), stop the compressor and observe the gauges. If the suction pressure increases to 10 or 15 psig (169.69 or 204.04 kPa), start the compressor and pump the system down to 1 or 2 psig again and stop the compressor. The suction pressure should remain at around this pressure. If not, repeat the pump-down process. If the pump-down process is repeated more than three times, the compressor discharge valves may be leaking. In this case the discharge service valve must be closed to prevent refrigerant bleeding into the low side of the system. *Caution:* Do not start the compressor when the discharge service valve is closed.
5. On units equipped with a low-pressure control set at higher pressure than 1 or 2 psig, it will be necessary to electrically bypass the control to keep the compressor running while pumping the system down.
6. Relieve any remaining pressure on the low side of the system by opening the low side hand valve on the gauge manifold. Do not attempt to weld or solder on a system having refrigerant pressure inside.
7. The necessary repairs can now be made. It is desirable to install a new liquid line drier before recharging the system.
8. To put the system back in operation, open the compressor discharge service valve and open the liquid line (king) valve. Allow a small amount of refrigerant to escape through the gauge manifold; then close the gauge manifold low side hand valve.
9. Start the unit and check the refrigerant charge. Add any required refrigerant to bring the system to full charge.

4.5 PROCEDURE FOR PUMPING A SYSTEM OUT

Pumping a system out is the process used to save the refrigerant in a system that does not have service valves, when repairs are to be made. This procedure is also used when repairs are to be made to the high side of the system and a system pump-down cannot be done. To accomplish this process, a portable condensing unit and a clean, dry refrigerant cylinder are required. The portable condensing unit is used to pump the refrigerant from the system and discharge it into the refrigerant cylinder where it is charged back into the system.

To pump a system out, use the following procedure:

1. Stop the unit.
2. Connect the gauge manifold to the system (see Figure 4-4).
3. Connect the center hose on the gauge manifold to the suction service valve on the portable condensing unit.
4. Connect the liquid line connection on the portable condensing unit to the valve on the refrigerant cylinder. Leave this connection loose for purging air from the lines and from the portable condensing unit. See Figure 4-5.

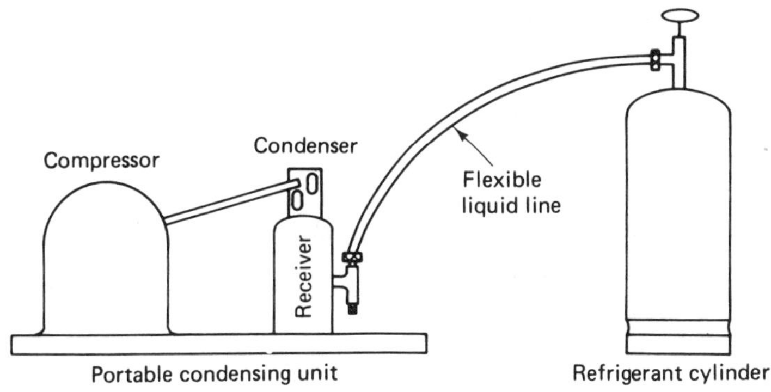

Figure 4–5 Connections from portable condensing unit to refrigerant cylinder.

5. Slowly crack the system service valves off the back seat until pressure is indicated on the gauges. Open the service valves on the portable condensing unit.
6. Open the hand valves on the gauge manifold and allow the refrigerant to blow out the connection on the refrigerant cylinder for a few seconds. Tighten the hose connection on the refrigerant cyclinder.
7. Open the refrigerant cylinder valve.
8. Start the portable condensing unit and pump the refrigerant from the system and into the cylinder. To prevent overloading the portable

condensing unit, regulate the suction pressure by partially closing the hand valves on the gauge manifold. *Caution:* Be sure not to overcharge the cylinder with refrigerant. Use as many cylinders as required by weight.

9. Remove the refrigerant until only 1 or 2 psig (108.17 or 114.783 kPa) is left inside the system. Close off all valves to prevent the refrigerant from escaping from the cylinder and the portable condensing unit.
10. Purge any remaining pressure from the refrigeration system and make the required repairs. It is desirable to install a new liquid line drier before charging the system with refrigerant.
11. To put the system back in operation, it should be completely evacuated. Then charge the refrigerant from the cylinder and the portable condensing unit back into the system. It may be desirable to install a drier in the charging line to remove any contaminants from the system. See Figure 4-6.

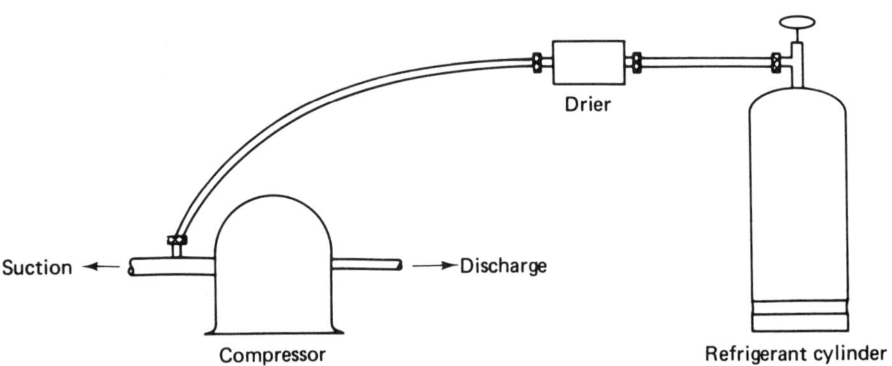

Figure 4–6 Liquid drier in charging line.

4.6 PROCEDURE FOR LEAK TESTING

When the refrigerant has escaped from the system, the leak must be found and repaired or the refrigerant will escape again. Refrigeration systems must be gas-tight for two reasons: first, any leakage will result in a loss of the refrigerant charge; second, leaks will allow air and moisture to enter the system when the pressure is reduced below 0 psig (100.989 kPa). Refrigerant leaks can and do occur at any time and at any point due to low-quality workmanship, age of the equipment, and vibration.

Because leak detection is a common service procedure, the service technician must be familiar with the various methods used. Each method has its advantages and disadvantages, depending on the circumstances surrounding the system. All the preferred methods require that pressure be introduced into the system. The three most popular methods are: electronic leak detection, halide-torch leak detection, and soap-bubble leak detection. However, the electronic method is not very satisfactory in a heavy concentration of refrigerant because it is extremely sensitive; and the halide torch is not generally satisfactory when very small leaks or blowing winds are encountered. Under most other conditions these two tests are found most satisfactory. Because electronic leak detectors are very sensitive, the soap-bubble test is more satisfactory for finding small leaks in a suspected area. Sometimes more than one of these methods may be used to locate one leak.

Evacuation is not a recommended leak detection method for many reasons. First, evacuation will not pinpoint a leak. Second, a flake of paint or a grain of sand may cover the leak during evacuation, preventing the loss of vacuum in the system. Third, more air and moisture are drawn into the system requiring more evacuation for their removal. Therefore, one of the above three methods is preferred.

To leak-test a system, use the following procedure:

1. Stop the unit. This is to reduce the air movement as much as possible.
2. Choose the leak detection method most suitable for the situation at hand.
3. Connect the gauge manifold to the refrigeration system (see Figure 4-4).
4. Be sure that the system has at least 35 psig pressure (341.44 kPa) inside it. If not, increase the pressure by adding refrigerant vapor to the system. Be sure to use the same type of refrigerant as that in the system.
5. Test each joint, fitting, and gasket in the entire system. Mark each leak as it is found. The repairs can be made after all the leaks have been located. Any joints or areas that have signs of oil residue on them should be given special attention. If no leak is found with the halide torch or the electronic leak detector, use a soap-bubble solution to be sure no leaks exist.

6. Either pump the system down, pump the system out, or purge the refrigerant from the system, depending on the amount of charge in the unit. Usually a refrigerant charge of less than 10 lb. (4.5359 Kg) is purged. A charge of more than 10 lb. (4.5359 Kg) is generally saved.
7. Repair the leaks and test to be sure that all leaks have been stopped.
8. Evacuate the system.
9. Install new liquid line driers on the system.
10. Recharge the system with the proper amount and type of clean, dry refrigerant.

4.7 PROCEDURE FOR EVACUATING A SYSTEM

Evacuation is the process used to remove air and moisture from a refrigeration system. Evacuation is accomplished by use of pumps especially designed for this purpose. A discarded refrigeration compressor is not suitable. Never run a motor compressor while the system is under a vacuum. To do so may result in serious damage to the motor winding. Evacuation is required any time a system has been contaminated or the compressor or system has been exposed to the atmosphere for long periods of time.

Purging a system will remove a good portion of the air, and driers will remove a part of any moisture from the system—but only up to the capacity of the drier. Therefore there are still some contaminants left in the system, and evacuation is the best means of being reasonably sure that the system is free of these contaminants.

There are basically two evacuation procedures: (1) simple evacuation, and (2) triple evacuation. Simple evacuation is used on systems containing only a minimum of contaminants. Triple evacuation is used on systems containing a greater amount of contaminants.

To use the simple evacuation method, use the following procedures:

1. Connect the gauge manifold to the system.
2. Purge all pressure from the refrigeration system by opening the system service valves and the gauge manifold hand valves.
3. Connect the center hose on the gauge manifold to a vacuum pump (see Figure 4-7).
4. Start the vacuum pump and pump a vacuum of at least 1,000 microns. A vacuum of 500 is preferable.
5. Close off the gauge manifold hand valves.

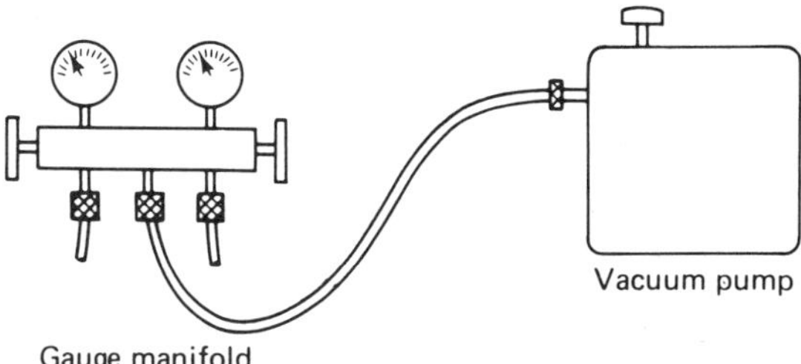

Figure 4–7 Gauge manifold connected to vacuum pump.

6. Stop the vacuum pump. Do not stop the vacuum pump before closing the gauge manifold hand valves. This is to prevent air being drawn into the system.
7. Disconnect the center hose of the gauge manifold from the vacuum pump and connect it to a cylinder containing the proper type of refrigerant (see Figure 4-8).

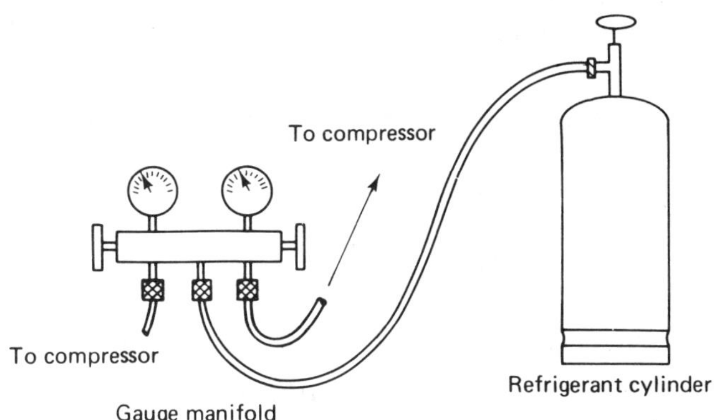

Figure 4–8 Gauge manifold connected to refrigerant cylinder.

8. Open the cylinder valve.
9. Loosen the center hose connection at the gauge manifold. Purge the hose for a few seconds, then tighten the connection.
10. Open the gauge manifold hand valves and admit refrigerant into the system.
11. Close the high side hand valve on the gauge manifold.
12. Start the unit and add the proper charge of refrigerant.

To use the triple-evacuation method, use the following procedure:

1. Connect the gauge manifold to the system (see Figure 4-4).
2. Purge all pressure from the system by opening the system service valves and the gauge manifold hand valves.
3. Connect the center hose on the gauge manifold to the vacuum pump (see Figure 4-8).
4. Start the vacuum pump. Do not stop the vacuum pump before closing the gauge manifold hand valves. This is to prevent air being drawn into the system.
7. Disconnect the center hose of the gauge manifold from the vacuum pump and connect it to a cylinder containing the proper type of refrigerant.
8. Open the cylinder valve.
9. Loosen the center hose connection at the gauge manifold. Purge the hose for a few seconds, then tighten the connection.
10. Open the gauge manifold hand valves and admit refrigerant into the system until a pressure of about 5 psig (135.34 kPa) is indicated on the gauges.
11. Close the refrigerant cylinder valve and the gauge manifold hand valves.
12. Disconnect the hose from the cylinder.
13. Open the gauge manifold hand valves and purge the pressure from the system.
14. Repeat steps 3 through 13.
15. Repeat steps 3 through 9 only. Pump a vacuum of 500 microns rather than 1,500 microns.
16. Open the gauge manifold hand valves and admit refrigerant into the system until cylinder pressure is indicated on the gauges.
17. Close the high side gauge manifold hand valve.
18. Start the unit and add the proper charge of refrigerant.

4.8 PROCEDURE FOR CHARGING REFRIGERANT INTO A SYSTEM

The performance of an air conditioning or refrigeration system is highly dependent on the proper charge of refrigerant in the system. A system that is undercharged will operate with a starved evaporator. Low suction pressures, loss of system capacity, and possible compressor overheating are the results of an undercharged system. On the other hand, an overcharge of refrigerant will back up in the condenser. High discharge pressures and liquid refrigerant flooding of the compressor with possible compressor damage are the results of an overcharged system. Larger systems can tolerate a reasonable amount of overcharging or undercharging without severe effects, but some of the smaller systems have a critical charge. The system must be properly charged to realize proper operation.

The amount of charge will depend on the size of the system in btu, the length of the lines, the type of refrigerant, and the operating temperature. Therefore, each system must be considered separately. The unit nameplate will usually indicate what type and the approximate weight of the refrigerant.

There are two methods of charging refrigerant into a system: (1) liquid charging, and (2) vapor charging. Liquid charging is much faster than vapor charging and is, therefore, used extensively on large field built-up systems. Never charge liquid refrigerant into the compressor suction or discharge service valves. Vapor charging is normally done only when small amounts of refrigerant are to be added to a system. Vapor charging also allows the refrigerant to be charged into the compressor suction service valve.

1. Connect the gauge manifold to the system (see Figure 4-4).
2. Open the system service valves and purge air from the lines.
3. Connect a charging line to the liquid line (king) valve. This line should include a liquid drier to prevent contaminants from entering the system.
4. Connect the charging line to the liquid valve on a cylinder of the proper type refrigerant. See Figure 4-9.

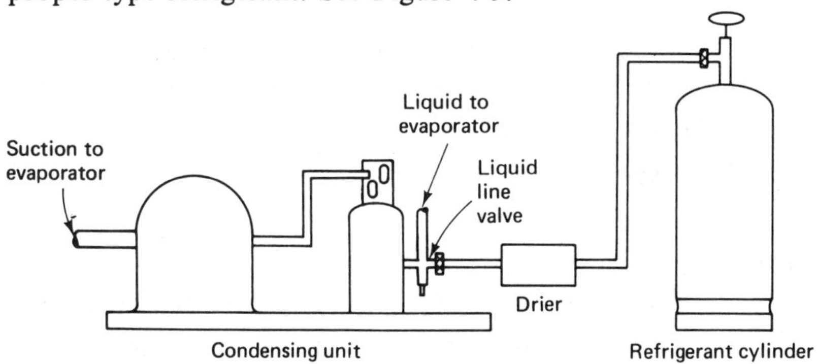

Figure 4–9 Connections for liquid charging a system.

5. Open the refrigerant cylinder liquid valve.
6. Loosen the charging line connections at the liquid line (king) valve and allow the refrigerant to escape for a few seconds.
7. If the correct charge weight is known, the refrigerant cylinder can be weighed to see when sufficient refrigerant has been charged into the system.
8. Close the liquid line (king) valve and start the compressor. Allow the liquid refrigerant to enter the liquid line until the correct weight has been charged into the system. See Step 10.
9. If the correct charge weight is not known, the liquid line valve must be opened periodically and the system operation observed. If more refrigerant is needed, close the liquid line valve again and charge more refrigerant into the system. Repeat this process until the proper charge is indicated. See Step 10.
10. Closely watch the discharge pressure gauge. A sudden increase in discharge pressure indicates that the capacity of the condenser and receiver has been reached. Stop charging the unit immediately and open the liquid line valve. Any additional refrigerant must be charged into the system by the vapor method.

To use the vapor-charging method, use the following procedure:

1. Connect the gauge manifold to the system (see Figure 4-4).
2. Connect the center line on the gauge manifold to a cylinder of the proper type of refrigerant (see Figure 4-8).
3. Open the refrigerant cylinder valve and hand valves on the gauge manifold.
4. Loosen the hose connection at the system service valves and allow the refrigerant to escape for a few seconds. Retighten the connections.
5. Close the gauge manifold hand valves.
6. Crack the system service valves.
7. Start the unit.
8. Open the low side gauge manifold hand valve and charge refrigerant into the system until the proper charge has been added.
9. Closely watch the discharge pressure gauge during the charging process to be certain that the system is not overcharged.

4.9 PROCEDURE FOR DETERMINING THE PROPER REFRIGERANT CHARGE

Determining the proper refrigerant charge is an important step in servicing an air conditioning unit. A system that does not contain the proper charge of refrigerant will not operate to maximum efficiency. There are several ways to determine if a system is properly charged, such as: weighing the charge, using a sight glass, using a liquid level indicator, using the liquid subcooling method, using the superheat method, and using the manufacturer's charging charts.

To use the charge-by-weight method, use the following procedures:

1. Connect the gauge manifold to the system service valves (see Figure 4-4).
2. Purge all pressure from the system. Open both hand valves on the gauge manifold.
3. Connect the center line from the gauge manifold to a vacuum pump (see Figure 4-7).
4. Start the vacuum pump and pump a deep vacuum on the system.
5. Close the gauge manifold hand valves.
6. Stop the vacuum pump. Do not stop the vacuum pump before closing the gauge manifold hand valves. This is to prevent air entering the system.
7. Disconnect the center line from the vacuum pump and connect it to a cylinder of the proper type of refrigerant (see Figure 4-8). This cylinder may be a charging cylinder or a large cylinder placed on an accurate scale. Small systems require a charging cylinder, whereas large systems require a large cylinder.
8. Open the cylinder valve.
9. Loosen the center hose connection at the gauge manifold. Purge the hose for a few seconds, then tighten the connection.
10. Open both hand valves on the gauge manifold and charge refrigerant into the system. This must be done suddenly so that all the refrigerant will enter the system before the vacuum vanishes.
11. Start the unit and check the operation.

To use the sight-glass method, use the following procedure:

1. Start the unit and allow it to operate for several minutes.
2. Check the flow of refrigerant through the sight glass. A flashlight may be needed to see the flow adequately. See Figure 4-10.

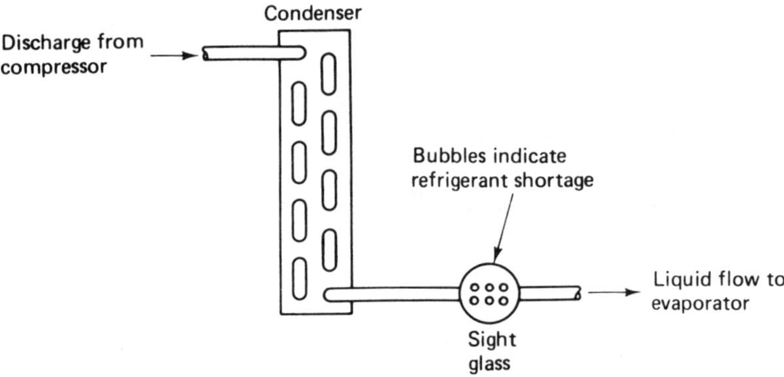

Figure 4–10 Checking refrigerant charge using a sight glass.

3. A steady stream of bubbles indicates the system is low on refrigerant. If these bubbles are intermittent, allow the system to operate a while longer to see if they will disappear. If the bubbles remain, the system is low on refrigerant.
4. Connect the gauge manifold to the system service valves (see Figure 4-4).
5. Connect the center line to a cylinder of the proper type of refrigerant (see Figure 4-8).
6. Crack the system services valves and open the gauge manifold hand valves, then loosen the connection on the refrigerant cylinder. Purge the lines for a few seconds. Tighten the connection.
7. Close the hand valves on the gauge manifold.
8. Open the refrigerant cylinder valve.
9. Open the low side hand valve on the gauge manifold and admit refrigerant into the system while observing the sight glass and the discharge pressure gauge. When the bubbles disappear, close the low side hand valve. Observe the sight glass. If the bubbles reappear, add more refrigerant into the system. Repeat this process until the bubbles do not reappear. A sudden increase in the discharge pressure indicates a system overcharge. Stop charging the unit and remove some of the refrigerant.

To use the liquid level indicator method, use the following procedure:

1. Start the unit and allow it to operate for 10 to 15 minutes.
2. After the system has been in operation long enough for the pressures to stabilize, crack the liquid level test port. A continuous flow of liquid refrigerant from the port indicates sufficient charge. A continuous flow of vapor from the port indicates a shortage of refrigerant. The test port is usually located in the lower section of the condenser. Some units may have a liquid level indicator in the receiver tank. See Figure 4-11.

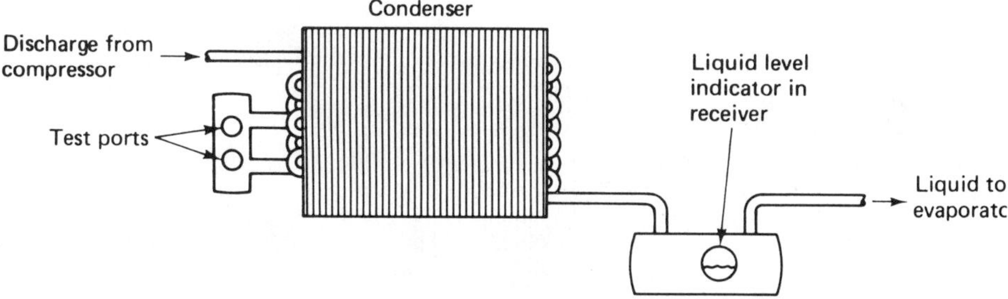

Figure 4–11 Liquid level indicator ports.

3. To add refrigerant to the system, connect the gauge manifold to the system service valves (see Figure 4-4). Close the gauge manifold hand valves. Connect the center charging line to a cylinder containing refrigerant of the proper type. Leave this connection loose (see Figure 4-8).
4. Crack the system service valves open until pressure is indicated on the gauges.
5. Open the gauge manifold hand valves. Allow refrigerant to escape from the hose connection on the cylinder for a few seconds, then tighten the connection.
6. Close the high side gauge manifold hand valve.
7. Open the refrigerant cylinder valve and charge refrigerant into the system.
8. Periodically open the liquid level test port and check for liquid refrigerant.
9. Continue to add refrigerant until a continuous stream of liquid refrigerant is obtained.

10. When liquid is obtained, close the low side gauge manifold hand valves. Allow the unit to operate a few minutes and check for liquid again. If a continuous stream of liquid is not indicated, add more refrigerant and check again. Continue this process until the proper charge is in the system.

To use the liquid subcooling method, use the following procedure:

1. Start the unit and allow it to operate for 10 to 15 minutes.
2. Strap a thermometer to the liquid line at the condenser outlet. See Figure 4-12.

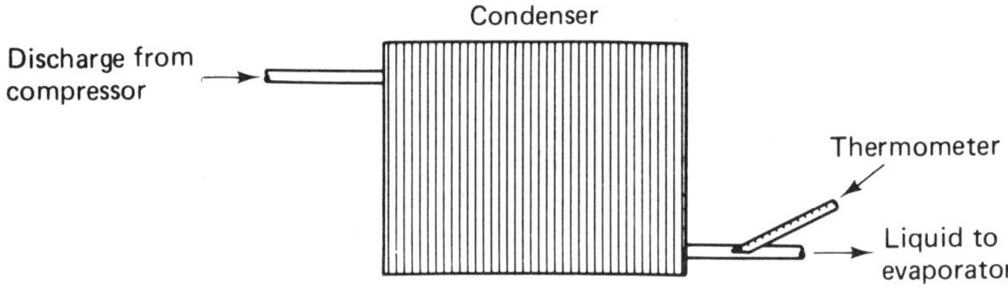

Figure 4–12 Checking liquid subcooling during charging process.

3. After the system has operated long enough for the pressures to stabilize, check the liquid temperature leaving the condenser as indicated on the thermometer.
4. Connect the gauge manifold to the system service valves. Close the gauge manifold hand valves (see Figure 4-4).
5. Open the system service valves until pressure is indicated on the gauges.
6. Loosen the hose connections on the gauge manifold and allow refrigerant vapor to escape for a few seconds, then retighten the connections.
7. Compare the liquid line temperature to the condensing temperature. The condensing temperature is determined by relating the discharge pressure to the corresponding temperature on a pressure-temperature chart. The liquid line temperature is indicated on the thermometer installed in step 2 above. The liquid line temperature should be approximately 5° F (2.35° C) below the condensing temperature. If less than 5° F (2.35° C), more refrigerant is needed in the system. See Table 4-1.

Table 4–1 Temperature-Pressure Chart (Table 4–1 TS, Courtesy of Alco Controls Division, Emerson Electric).

°F	R-12	R-13	R-22	R-500	R-502	R-717 Ammonia	°F	R-12	R-13	R-22	R-500	R-502	R-717 Ammonia
−100	***27.0***	7.5	***25.0***	***26.4***	***23.3***	***27.4***	16	18.4	211.9	38.7	24.1	47.7	29.4
−95	***26.4***	10.9	***24.1***	***25.7***	***22.1***	***26.8***	18	19.7	218.8	40.9	25.7	50.1	31.4
−90	***25.8***	14.2	***23.0***	***24.9***	***20.7***	***26.1***	20	21.0	225.7	43.0	27.3	52.5	33.5
−85	***25.0***	18.2	***21.7***	***24.0***	***19.0***	***25.3***	22	22.4	233.0	45.3	28.9	54.9	35.7
−80	***24.1***	22.3	***20.2***	***22.9***	***17.1***	***24.3***	24	23.9	240.3	47.6	30.6	57.4	37.9
−75	***23.0***	27.1	***18.5***	***21.7***	***15.0***	***23.2***	26	25.4	247.8	49.9	32.4	60.0	40.2
−70	***21.9***	32.0	***16.6***	***20.3***	***12.6***	***21.9***	28	26.9	255.5	52.4	34.2	62.7	42.6
−65	***20.5***	37.7	***14.4***	***18.8***	***10.0***	***20.4***	30	28.5	263.2	54.9	36.0	65.4	45.0
−60	***19.0***	43.5	***12.0***	***17.0***	***7.0***	***18.6***	32	30.1	271.3	57.5	37.9	68.2	47.6
−55	***17.3***	50.0	***9.2***	***15.0***	***3.6***	***16.6***	34	31.7	279.5	60.1	39.9	71.1	50.2
−50	***15.4***	57.0	***6.2***	***12.8***	0.0	***14.3***	36	33.4	287.8	62.8	41.9	74.1	52.9
−45	***13.3***	64.6	***2.7***	***10.4***	2.1	***11.7***	38	35.2	296.3	65.6	43.9	77.1	55.7
−40	***11.0***	72.7	0.5	***7.6***	4.3	***8.7***	40	37.0	304.9	68.5	46.1	80.2	58.6
−35	***8.4***	81.5	2.6	***4.6***	6.7	***5.4***	45	41.7	327.5	76.0	51.6	88.3	66.3
−30	***5.5***	90.9	4.9	***1.2***	9.4	***1.6***	50	46.7	351.2	84.0	57.6	96.9	74.5
−28	***4.3***	94.9	5.9	0.1	10.5	0.0	55	52.0	376.1	92.6	63.9	106.0	83.4
−26	***3.0***	98.9	6.9	0.9	11.7	0.8	60	57.7	402.3	101.6	70.6	115.6	92.9
−24	***1.6***	103.0	7.9	1.6	13.0	1.7	65	63.8	429.8	111.2	77.8	125.8	103.1
−22	***0.3***	107.3	9.0	2.4	14.2	2.6	70	70.2	458.7	121.4	85.4	136.6	114.1
−20	0.6	111.7	10.2	3.2	15.5	3.6	75	77.0	489.0	132.2	93.5	148.0	125.8
−18	1.3	116.2	11.3	4.1	16.9	4.6	80	84.2	520.8	143.6	102.0	159.9	138.3
−16	2.1	120.8	12.5	5.0	18.3	5.6	85	91.8	—	155.7	111.0	172.5	151.7
−14	2.8	125.7	13.8	5.9	19.7	6.7	90	99.8	—	168.4	120.6	185.8	165.9
−12	3.7	130.5	15.1	6.8	21.2	7.9	95	108.3	—	181.8	130.6	199.7	181.1
−10	4.5	135.4	16.5	7.8	22.8	9.0	100	117.2	—	195.9	141.2	214.4	197.2
−8	5.4	140.5	17.9	8.8	24.4	10.3	105	126.6	—	210.8	152.4	229.7	214.2
−6	6.3	145.7	19.3	9.9	26.0	11.6	110	136.4	—	226.4	164.1	245.8	232.3
−4	7.2	151.1	20.8	11.0	27.7	12.9	115	146.8	—	242.7	176.5	262.6	251.5
−2	8.2	156.5	22.4	12.1	29.4	14.3	120	157.7	—	259.9	189.4	280.3	271.7
0	9.2	162.1	24.0	13.3	31.2	15.7	125	169.1	—	277.9	203.0	298.7	293.1
2	10.2	167.9	25.6	14.5	33.1	17.2	130	181.0	—	296.8	217.2	318.0	315.0
4	11.2	173.7	27.3	15.7	35.0	18.8	135	193.5	—	316.6	232.1	338.1	335.0
6	12.3	179.8	29.1	17.0	37.0	20.4	140	206.6	—	337.3	247.7	359.1	365.0
8	13.5	185.9	30.9	18.4	39.0	22.1	145	220.3	—	358.9	266.1	381.1	390.0
10	14.6	192.1	32.8	19.7	41.1	23.8	150	234.6	—	381.5	281.1	403.9	420.0
12	15.8	198.6	34.7	21.2	43.2	25.6	155	249.5	—	405.1	298.9	427.8	450.0
14	17.1	205.2	36.7	22.6	45.5	27.5	160	265.1	—	429.8	317.4	452.6	490.0

BOLD ITALIC FIGURES – VACUUM, UPRIGHT FIGURES – PRESSURE
[1]ALCO CO.NTROLS DIVISION, EMERSON ELECTRIC CO, P.O. BOX 12700, ST. LOUIS, MISSOURI 63141

8. To add refrigerant, connect the center charging hose on the gauge manifold to the valve on a cylinder containing the proper type of refrigerant. Do not tighten this connection (see Figure 4-8).
9. Open the hand valves on the gauge manifold. Allow the refrigerant to escape a few seconds, then tighten the connection on the cylinder valve.
10. Close the high side hand valve on the gauge manifold.
11. Open the cylinder valve and charge refrigerant into the system.
12. Observe the thermometer on the liquid line and the discharge temperature. When the desired subcooling is obtained, close the low side hand valve on the gauge manifold to stop adding refrigerant to the system.
13. Continue operation of the unit for a few minutes to allow the pressures to stabilize, and check the amount of subcooling. If additional subcooling is required, add more refrigerant to the system. Continue this process until the 5° F (2.35° C) subcooling remains stable.

To use the superheat method, use the following procedure:

1. Start the unit and allow it to operate for 10 to 15 minutes.
2. Strap a thermometer on the suction line about 6 in. from the compressor. Insulate the thermometer bulb so that accurate readings are indicated. See Figure 4-13.

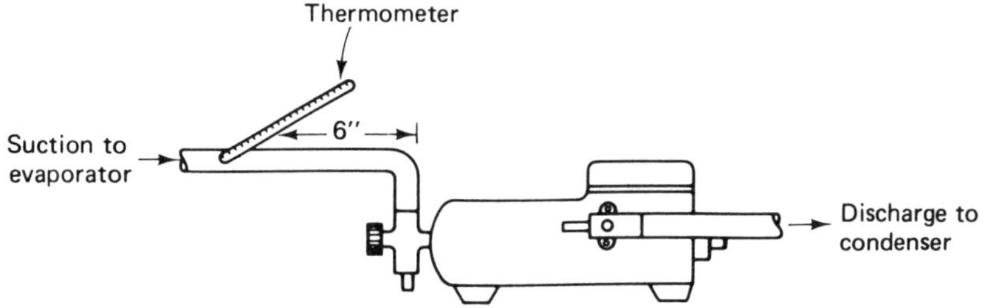

Figure 4–13 Thermometer strapped to suction line.

3. If a low side pressure port is available, connect the low side gauge to the system by connecting the low side hose on the gauge manifold to the pressure port. If a pressure port is not available, strap a thermometer to a return bend on the evaporator. Do not put the thermometer on a fin. Insulate the thermometer bulb so that accurate readings may be indicated. See Figure 4-14.

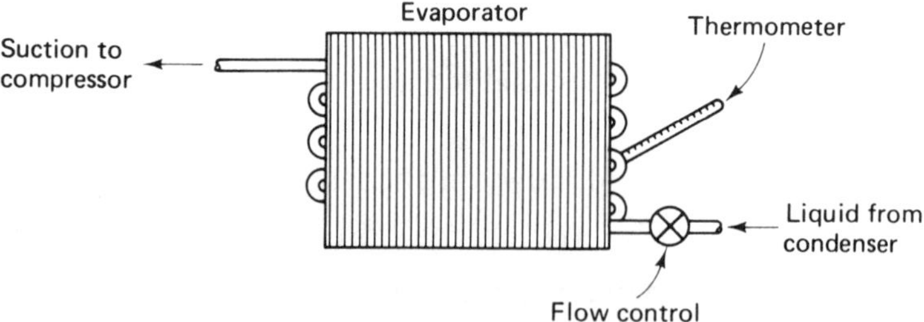

Figure 4–14 Thermometer strapped to evaporator return bend.

4. If a pressure port is available, determine the difference between the suction line temperature and the saturation temperature equivalent to the suction pressure with the unit running. The saturation temperature is determined by relating the suction pressure to the

corresponding temperature on a pressure-temperature chart. The temperature difference should be approximately 20° F to 30° F (11.11 to 16.67° C). See Table 4-1.

If no pressure port is available, check the difference between the two thermometers. The difference in temperature should be approximately 20° F to 30° F (11.11° C to 16.67° C). When the unit is operating at a normal operating condition, a superheat of 20° F to 30° F (-6.6° C to -1.1° C) is satisfactory. A superheat lower than approximately 20° F (-6.6° C) indicates an overcharge of refrigerant. A superheat higher than approximately 30° F (-1.1° C) indicates an undercharge of refrigerant.

5. To add refrigerant, connect the low side gauge of the gauge manifold to the pressure port of the system. If a low side pressure port is not available, a saddle valve must be installed. See Figure 4-15.

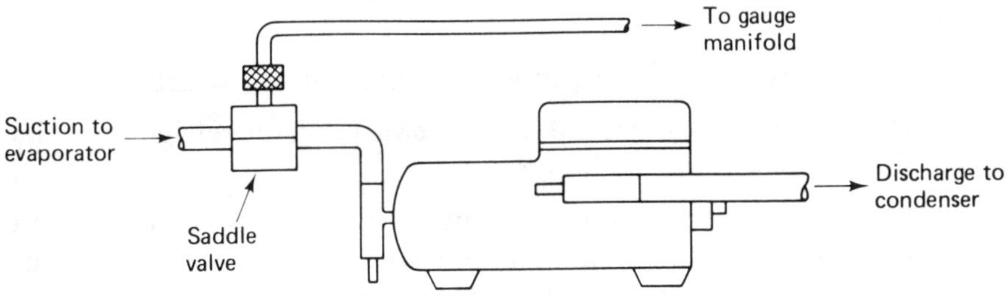

Figure 4–15 Installation of saddle valve.

6. Connect the center charging hose on the gauge manifold to the valve on a cylinder of the proper type refrigerant. Do not tighten this connection (see Figure 4-8).
7. Open the gauge manifold hand valve and allow the refrigerant to escape for a few seconds. Tighten the connection on the cylinder valve.
8. Open the refrigerant cylinder valve and charge refrigerant into the system.
9. Observe the superheat by the method used above in step 4. When the proper superheat is reached, stop charging refrigerant into the system.
10. Allow the unit to continue running until the pressures and temperatures have stabilized. If more refrigerant is needed, repeat the charging process until the desired superheat is stabilized.

To use the manufacturer's charging chart method, use the following procedure:

1. Start the unit and allow it to operate for 10 to 15 minutes.
2. Connect the gauge manifold to the system service valves (see Figure 4-4).
3. Connect the center charging hose on the gauge manifold to the valve on a cylinder of the proper type refrigerant. Do not tighten this connection.
4. Open the hand valves on the gauge manifold and allow refrigerant to escape from the loose connection for a few seconds. Tighten the connection on the cylinder valve.
5. Close the high side hand valve on the gauge manifold.
6. Obtain a copy of the manufacturer's charging chart for that model unit.
7. Compare the pressures to those indicated on the chart.
8. If more refrigerant is needed, open the valve on the refrigerant cylinder and add refrigerant to the system.
9. Stop charging after a few minutes by closing the low side hand valve on the gauge manifold and compare the pressure readings to those indicated on the chart. Repeat this procedure until the system pressures compare with those indicated on the chart.

4.10 PROCEDURE FOR DETERMINING THE COMPRESSOR OIL LEVEL

All refrigeration compressors require a specified amount of oil. This oil is required for lubrication of the moving parts, and helps to make a refrigerant seal between the components. An abnormally low oil level will probably result in a loss of lubrication and compressor damage. An excess of lubricating oil will result in oil slugging, probable damage to the compressor valves, and lost system efficiency due to oil logging of the evaporator.

To use the sight-glass method, use the following procedure:

1. Start the unit and allow it to operate for 10 to 15 minutes.
2. Check the oil level in the sight glass. See Figure 4-16.

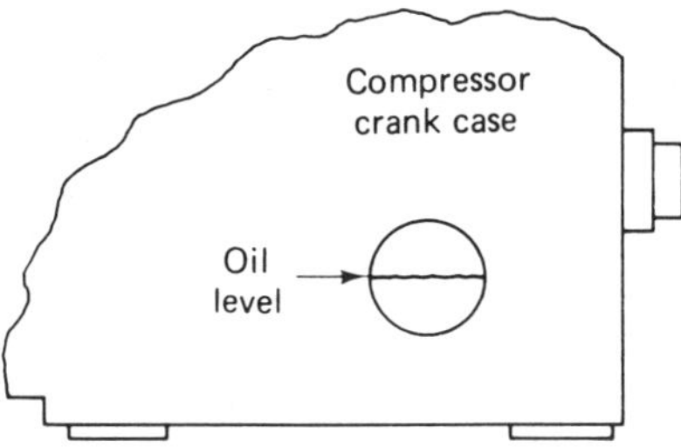

Figure 4–16 Checking oil level in sight glass.

A flashlight may be needed to determine the level accurately. The oil level should be at or slightly above the center of the sight glass while the unit is operating. If less than this, oil should be added. If more than this, the excess oil should be removed.

On sealed systems that do not have an oil sight glass, determining the amount of oil in the compressor is difficult. When a leak occurs and the amount of oil lost is small and can be reasonably calculated, add that amount to the system. If there has been a large amount of oil lost, the compressor must be removed, the oil drained, and the correct amount added to the compressor before placing it back in operation.

4.11 PROCEDURE FOR ADDING OIL TO A COMPRESSOR

Adding oil to a compressor is often required, and the service technician should be familiar with the different procedures used. There are three procedures used, depending on the type of system being repaired and the type of tools at hand. These procedures are: (1) the open system method, (2) the closed system method, and (3) the oil pump method.

To use the open system method, use the following procedures:

1. Connect the gauge manifold to the system service valves (see Figure 4-4).
2. Close the gauge manifold hand valves and open the compressor service valves.
3. Start the unit.

4. Front-seat the compressor suction service valve and run the unit until the low side pressure is reduced to 1 or 2 psig (107.86 kPa or 114.73 kPa). The lower pressure switch may need to be bypassed.
5. Stop the compressor.
6. Front-seat the compressor discharge service valve.
7. Open the low side hand valve on the gauge manifold and relieve the slight pressure remaining in the compressor.
8. Remove the oil fill plug and check the oil level with a dip stick that has been marked at the proper level. See Figure 4-17.

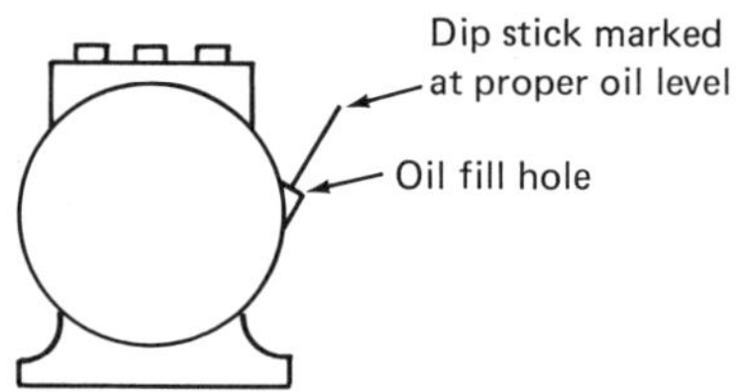

Figure 4–17 Checking oil level with a dip stick.

9. Pour the oil into the compressor crankcase until the proper level is reached. Use precaution to prevent contamination of the oil. See Figure 4-18.

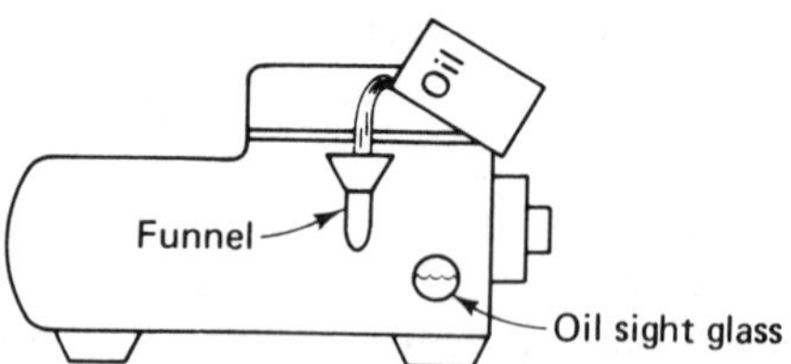

Figure 4–18 Pouring oil into a compressor.

10. Close the low side hand valve on the gauge manifold.
11. Slightly open the compressor suction service valve and allow a small amount of refrigerant to escape out of the oil fill hole.

12. Close off the compressor suction service valve.
13. Replace the oil fill plug and tighten.
14. Back-seat the compressor service valves.
15. Start the compressor and check the oil level in the compressor.
16. Remove the gauge manifold from the system.

To use the closed system method, use the following procedure:

1. Connect the gauge manifold to the system service valves (see Figure 4-4).
2. Place the center charging line in a container of clean dry oil. See Figure 4-19.

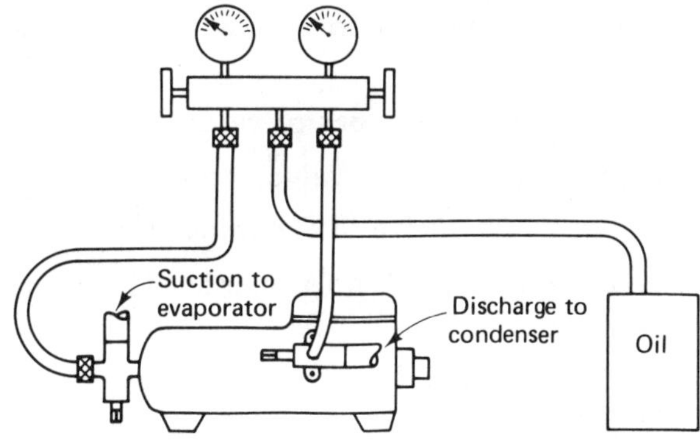

Figure 4–19 Adding oil to a closed system.

3. Open the system service valves until a pressure is indicated on the gauges.
4. Slightly open the low side hand valve on the gauge manifold and purge a small amount of refrigerant through the lines and the oil.
5. Front-seat the system service valve.
6. Start the unit and draw a vacuum on the compressor crankcase.
7. Open the low side hand valve on the gauge manifold and draw the oil into the compressor through the service valve. Be sure that the center line remains in the oil to prevent air being drawn into the system.
8. When a sufficient amount of oil has been drawn into the compressor, close the hand valve on the gauge manifold.

9. Back-seat the system service valve, and place the unit in normal operation.

To use the oil pump method, use the following procedure:

1. Connect the low side gauge of the gauge manifold to the system suction service valve.
2. Close the low side hand valve on the gauge manifold.
3. Crack the system service valve until pressure is indicated on the gauge.
4. Connect the center charging line to the oil pump. Do not tighten this connection.
5. Open the low side gauge manifold hand valve and purge refrigerant through the loose connection for a few seconds, then tighten the connection.
6. Place the oil pump in a container of clean dry oil.
7. Open the gauge manifold hand valve completely.
8. Move the system service valve to the midway position.
9. Pump oil into the system until the proper level is reached. See Figure 4-20.

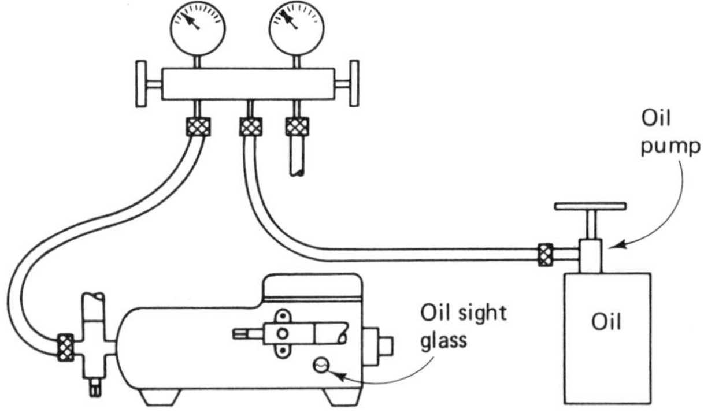

Figure 4–20 Adding oil to a system using an oil pump.

10. Back-seat the system service valve.
11. Close the low side hand valve on the gauge manifold.
12. Remove the gauge manifold and place the system in normal operation.

4.12 PROCEDURE FOR LOADING A CHARGING CYLINDER

Charging cylinders are ideal for charging systems that use less than 5 lb. (0.1417 Kg) of refrigerant. Charging cylinders are calibrated in ounces for each type of refrigerant. Therefore, it is possible to charge the exact amount of refrigerant into the system. The service technician should be familiar with the steps involved in loading a charging cylinder.

To load a charging cylinder, use the following procedure:

1. Connect the low pressure gauge of the gauge manifold to the charging cylinder. See Figure 4-21.

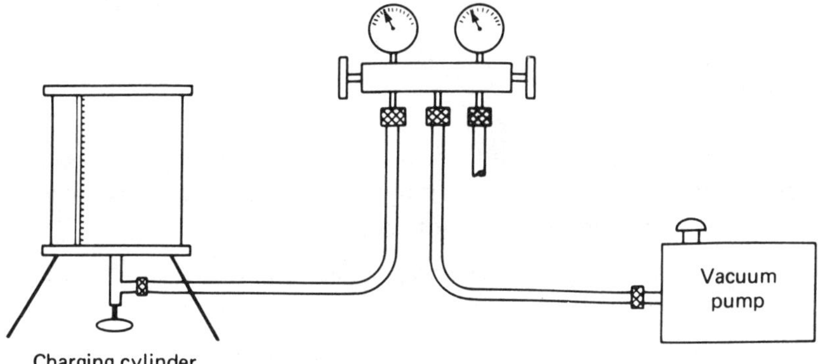

Figure 4–21 Connections for evacuating a charging cylinder.

2. Open the hand valve on the charging cylinder.
3. Open the low side hand valve on the gauge manifold and purge all pressure from the charging cylinder.
4. Connect the center charging line on the gauge manifold to the vacuum pump.
5. Start the vacuum pump and draw as deep a vacuum as possible with the vacuum pump.
6. Close the low side hand valve on the gauge manifold.
7. Remove the center charging line from the vacuum pump and connect it to the valve on a cylinder of the proper type of refrigerant. If there is a liquid valve, connect the line to it. See Figure 4-22.
8. Open the refrigerant cylinder valve and loosen the center charging line connection at the gauge manifold. Allow the refrigerant to escape for a few seconds, then retighten the connection.

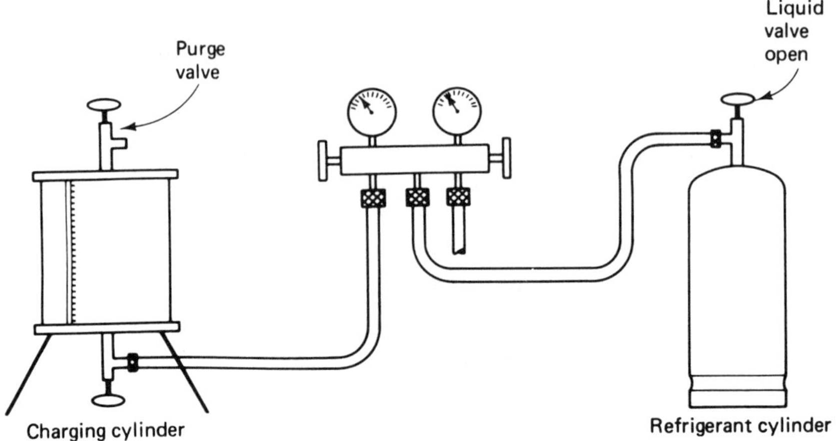

Figure 4–22 Line connections for loading a charging cylinder.

9. Invert the refrigerant cylinder, open the liquid valve, so that liquid refrigerant will enter the charging line. If the line is connected to a liquid valve, the cylinder is not to be inverted.
10. Open the low side hand valve on the gauge manifold and allow liquid refrigerant to be drawn into the charging cylinder.
11. When the desired amount of refrigerant has been drawn into the charging cylinder, close the refrigerant cylinder valve. If the refrigerant stops flowing before the desired amount is in the charging cylinder, open the vent valve on top of the charging cylinder to permit the escape of vapor, thus allowing more liquid to enter (see Figure 4-22).
12. Close the charging cylinder hand valve and disconnect the lines from the charging cylinder. Be sure that any liquid refrigerant in the lines does not come in contact with the skin or eyes.

4.13 PROCEDURE FOR CHECKING COMPRESSOR ELECTRICAL SYSTEMS

Potential Starting Relay System with Two-Terminal Overload

Potential (voltage) starting relays are nonpositional. The contacts are normally closed. These relays are generally used on single-phase compressors of ½ hp and larger. Use the following procedure for checking this type of system. Refer to Figure 4-23.

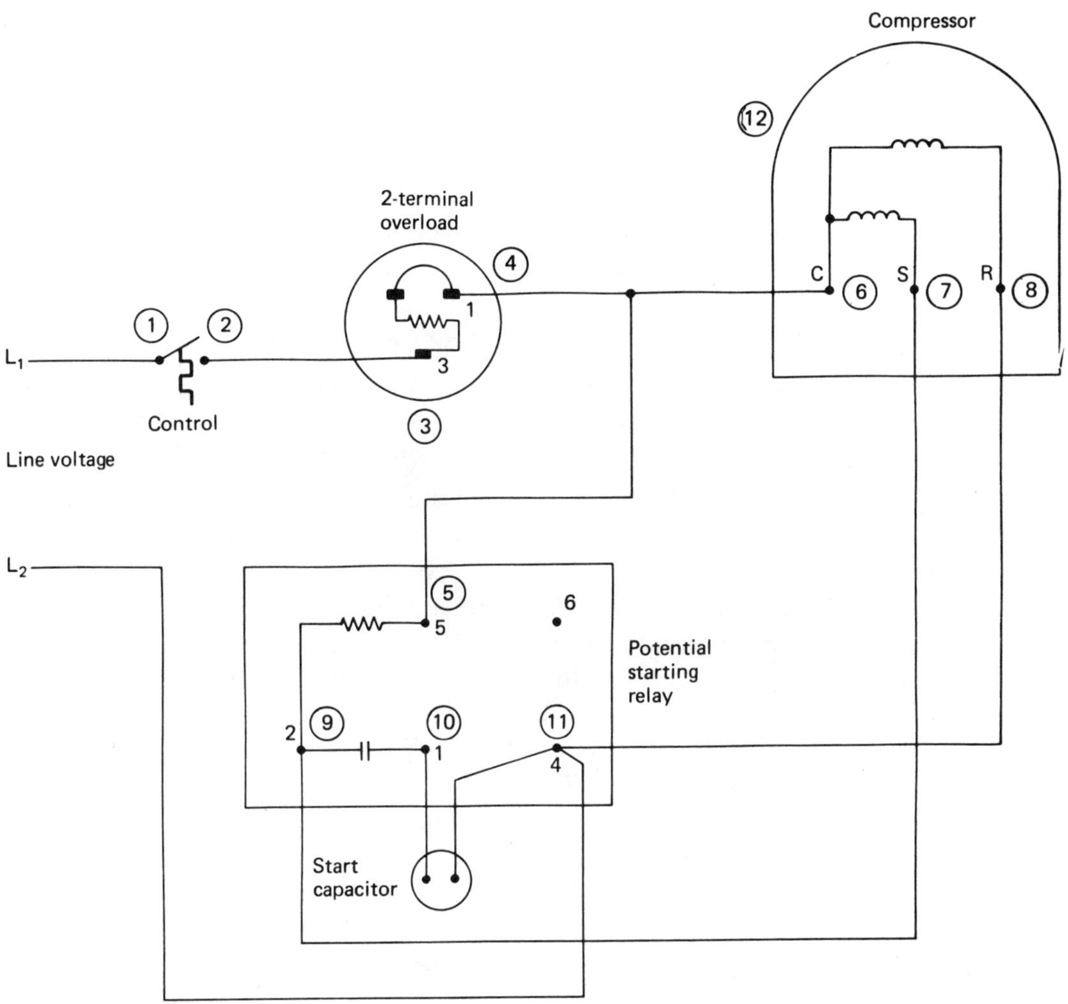

Figure 4–23 Potential starting relay system with a two-terminal overload.

1. Check to make certain that the proper voltage is being supplied to the unit. If proper voltage is supplied, continue to make the following checks. If the wrong voltage is supplied, correct the problem with the power supply.
2. Disconnect the electrical power from the unit.
3. If a fan motor is used, disconnect one of the leads.
4. Make the following tests with an ohmmeter. Be sure to zero the ohmmeter.

5. Check the continuity between points 1 and 2. Refer to Figure 4-23. If no continuity is found, the control contacts are open. Close the contacts by setting the control to demand operation. If continuity is found, continue to make the following checks. If no continuity is found with the control on demand, replace the control.
6. Check the continuity between points 3 and 4. If continuity is found, continue to make the following checks. If no continuity is found, wait about ten minutes and check the continuity again. The overload may be tripped. If no continuity is found now, replace the overload. Be sure to use an exact replacement.
7. Check the continuity between points 4 and 5. If continuity is found, continue to make the following checks. If no continuity is found, repair the broken wire or loose connection.
8. Check the continuity between points 4 and 6. If continuity is found, continue to make the following checks. If no continuity is found, repair the broken wire or loose connection.
9. Remove the wiring from the start terminal (point 7). Check the continuity between points 6 and 7. Compare this reading to that given by the manufacturer for that particular compressor motor. If proper continuity is found, continue to make the following checks. If no or improper continuity is found, replace the compressor motor. The start winding is defective.
10. Remove the wiring from the run terminal (point 8). Check the continuity between points 6 and 8. Compare this reading to that given by the manufacturer for that particular compressor motor. If proper continuity is found, continue to make the following checks. If no or improper continuity is found, replace the compressor motor. The run winding is defective.
11. Check the continuity between points 5 and 9. If continuity is found, continue to make the following checks. If no continuity is found, the start relay coil is open. Replace the relay. Be sure to use an exact replacement.
12. Check the continuity between points 9 and 10. If continuity is found, continue to make the following checks. If no continuity is found, the start relay contacts are defective. Replace the relay. Be sure to use an exact replacement.
13. Set the ohmmeter on R × 100,000. Check the continuity between points 10 and 11. If needle deflection is found, continue to make the following

checks. If no needle deflection is found, the capacitor is open. Replace the capacitor. Be sure to use an exact replacement.

14. Set the ohmmeter on r × 1. Check the continuity between points 10 and 11. If no continuity is found, continue to make the following checks. If continuity is found, the capacitor is shorted. Replace the capacitor. Be sure to use an exact replacement.
15. Check the continuity between points 6 and 12. If no continuity is found, continue to make the following checks. If continuity is found, the compressor motor is shorted. Replace the compressor.
16. Check the continuity between point 9 and the wire removed from point 7 (start terminal). If continuity is found, place the wire back on the terminal and continue to make the following checks. If no continuity is found, repair the wiring or loose connection.
17. Check the continuity between point 11 and the wire removed at point 8 (run terminal). If continuity is found, place the wire back on the terminal and continue to make the following checks. If no continuity is found, repair the wiring or loose connection.
18. If the above steps do not locate the trouble and the compressor will operate for a short period of time, disconnect the wire from point 10. Touch the wire to the same terminal and turn on the electricity. When the compressor starts, immediately remove the wire from the terminal. A slight spark may occur. If the compressor continues to operate, replace the start relay. The contacts are not opening. Be sure to use an exact replacement. If the compressor does not continue to run when the wire is removed, reconnect the wire and proceed to the next step.
19. Check the amperage draw through the wire to point 6 (common terminal) while the compressor is trying to run. Compare the amperage draw to the locked rotor (LR) amperage of the motor. If the amperage draw is very close to the LR listed by the manufacturer, replace the compressor. The compressor has mechanical problems.

Potential Starting Relay System with Three-Terminal Overload

Potential (voltage) starting relays are nonpositional. The contacts are normally closed. These relays are generally used on single-phase compressor motors of ½ hp and larger. Use the following procedure to check this type of system. See Figure 4-24.

1. Check to make sure that the proper voltage is being supplied to the unit. If the proper voltage is being supplied, continue to make the

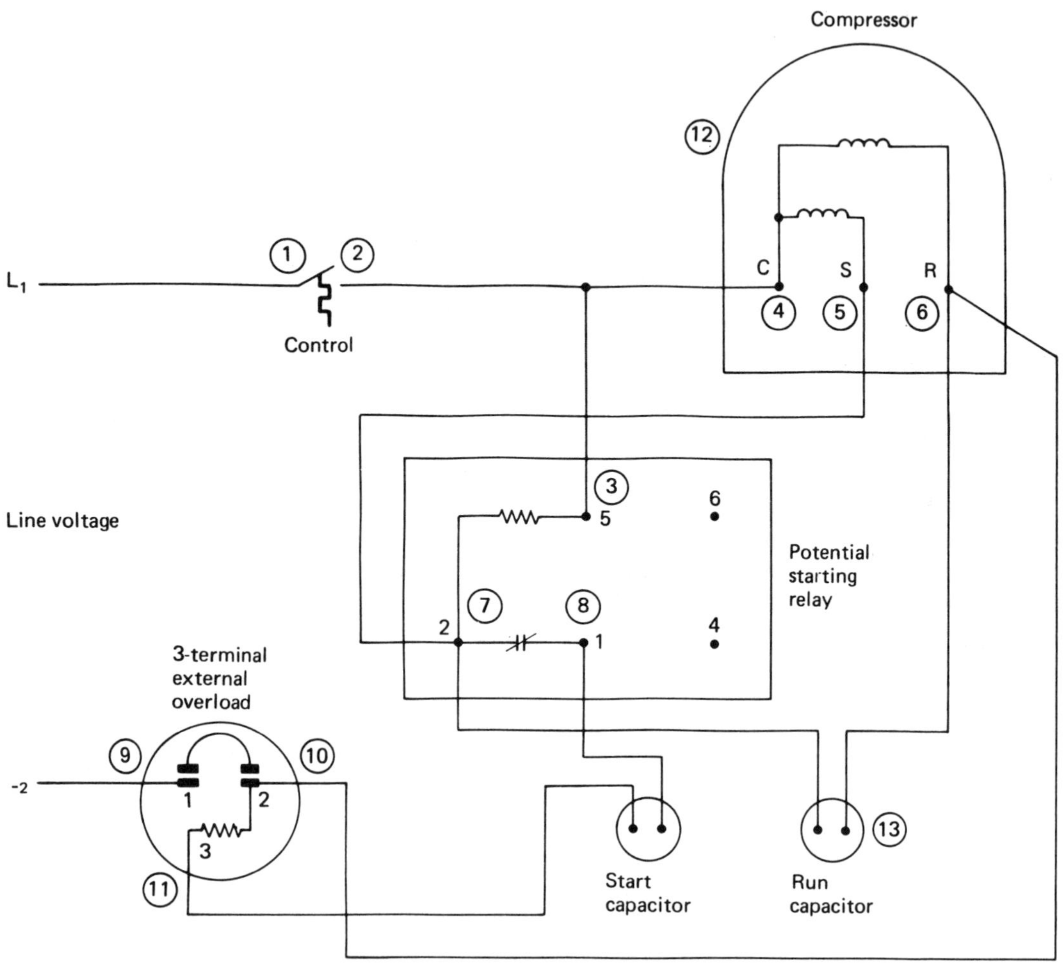

Figure 4–24 Potential starting relay system with three-terminal overload.

following checks. If the wrong voltage is being supplied, correct the problem with the power supply.

2. Disconnect the electrical power from the unit.
3. If a fan is used, disconnect one of the leads.
4. Make the following checks with an ohmmeter. Be sure to zero the ohmmeter.
5. Check the continuity between points 1 and 2. Refer to Figure 4-24. If continuity is found, continue to make the following checks. No

continuity indicates the control contacts are open. Close the contacts by setting the control to demand operation. If there is no continuity with the control on demand, replace the control.

6. Check the continuity between points 2 and 3. If continuity is found continue to make the following checks. If no continuity is found, repair the broken wire or loose connection.
7. Check the continuity between points 2 and 4. If continuity is found, continue to make the following checks. If no continuity is found, repair the broken wire or loose connection.
8. Remove the wiring from point 5 (start terminal). Check the continuity between points 4 and 5. Compare this reading to that given by the manufacturer for that particular compressor motor. If proper continuity is found, continue to make the following checks. If no or improper continuity is found, replace the compressor motor. The start winding is defective.
9. Remove the wiring from point 6 (run terminal). Check the continuity between points 4 and 6. Compare this reading to that given by the manufacturer for that particular compressor motor. If no or proper continuity is found, continue to make the following checks. If no or improper continuity is found, replace the compressor motor. The run winding is defective.
10. Check the continuity between points 3 and 7. If continuity is found, continue to make the following checks. If no continuity is found, replace the relay. The coil is faulty. Be sure to use an exact replacement.
11. Check the continuity between points 7 and 8. If continuity is found, continue to make the following checks. If no continuity is found, replace the relay. The contacts are faulty. Be sure to use an exact replacement.
12. Set the ohmmeter on R × 100,000. Check the continuity between points 8 and 11. If needle deflection is found, continue to make the following checks. If no needle deflection is found, the start capacitor is open. Replace the capacitor. Be sure to use an exact replacement.
13. Set the ohmmeter on R × 1. Check the continuity between points 9 and 11. If no continuity is found, continue to make the following checks. If continuity is found, the start capacitor is shorted. Replace the capacitor. Be sure to use an exact replacement.

14. Set the ohmmeter on R × 100,000. Check the continuity between points 7 and the wire removed from point 6 (run terminal). If needle deflection is found, continue to make the following checks. If no needle deflection is found, the run capacitor is open. Replace the capacitor. Be sure to use a proper replacement.
15. Set the ohmmeter on R × 1. Check the continuity between point 7 and the wire removed from point 6 (run terminal). If no continuity is found, continue to make the following checks. If continuity is found, the run capacitor is shorted. Replace the capacitor. Be sure to use a proper replacement.
16. Check the continuity between points 9 and 11. If continuity is found, continue to make the following checks. If no continuity is found, replace the overload. Be sure that the unit has been off for about ten minutes to allow the overload to cool. Be sure to use an exact replacement.
17. Check the continuity between points 4 and 12. If no continuity is found, continue to make the following checks. If continuity is found, replace the compressor. The motor is grounded to the case.
18. Check the continuity between point 7 and the wire removed from point 5 (start terminal). If continuity is found, continue to make the following checks. If no continuity is found, repair the wiring or loose connection. Reconnect the wiring to terminal (point 5).
19. Check the continuity between point 13 and the wire removed from point 6 (run terminal). If continuity is found, continue to make the following checks. If no continuity is found, repair the wiring or loose connection. Reconnect this wire to the terminal (point 6).
20. Check the continuity between point 10 and the wire removed from point 6 (run terminal). If continuity is found, continue to make the following checks. If no continuity is found, repair the wiring or loose connection. Reconnect this wire to terminal (point 6).
21. If the above steps do not indicate the trouble, and the compressor will operate for a short period of time, disconnect the wire from point 8. Touch the wire to the same terminal and turn on the electricity. When the compressor starts, immediately remove the wire from the terminal. A slight spark may occur. If the compressor continues to operate, replace the start relay. The contacts are not operating. Be sure to replace it with an exact replacement. If the compressor does not continue to operate when the wire is removed, reconnect the wire and proceed to the next step.

22. Check the amperage draw through the wire to point 4 (common terminal) while the compressor is trying to run. Compare the amperage draw to the locked rotor (LR) amperage of the motor. If the amperage draw is very close to the LR listed by the manufacturer, replace the compressor. The compressor has mechanical problems.

Current-type Starting Relay System with a Two-Terminal Overload

Current-type starting relays are positional-type relays. They must be mounted in the position indicated by the arrows on the relay. The contacts are normally open, and they are closed by an electromagnetic coil. They are opened by gravity, thus the purpose of mounting them in the proper position. These relays are generally used on fractional hp motors up to about ½ hp.

Use the following procedure to check this type of system. See Figure 4-25.

1. Check to make certain that the proper voltage is being supplied to the unit. If proper voltage is being supplied, continue to make the following checks. If the wrong voltage is being supplied, correct the problem with the power supply.
2. Disconnect the electrical power from the unit.
3. If a fan motor is being used, disconnect one of the leads.
4. Make the following checks with an ohmmeter. Be sure to zero the ohmmeter.
5. Check the continuity between points 1 and 2. Refer to Figure 4-25. No continuity indicates that the control contacts are open. Close the contacts by setting the control to demand operation. If continuity is found, continue to make the following checks. If no continuity is found with the control on demand, replace the control.
6. Check the continuity between points 3 and 4. If continuity is found, continue to make the following checks. If no continuity is found, wait about ten minutes and check the continuity again. The overload may be tripped. If no continuity now, replace the overload. Be sure to use an exact replacement.
7. Check the continuity between points 4 and 5. If continuity is found, continue to make the following checks. If no continuity is found, repair the wire or loose connections.
8. Remove the wire from point 6 (start terminal). Check the continuity between points 5 and 6. Compare this reading to that given by the manufacturer for that particular compressor motor. If proper

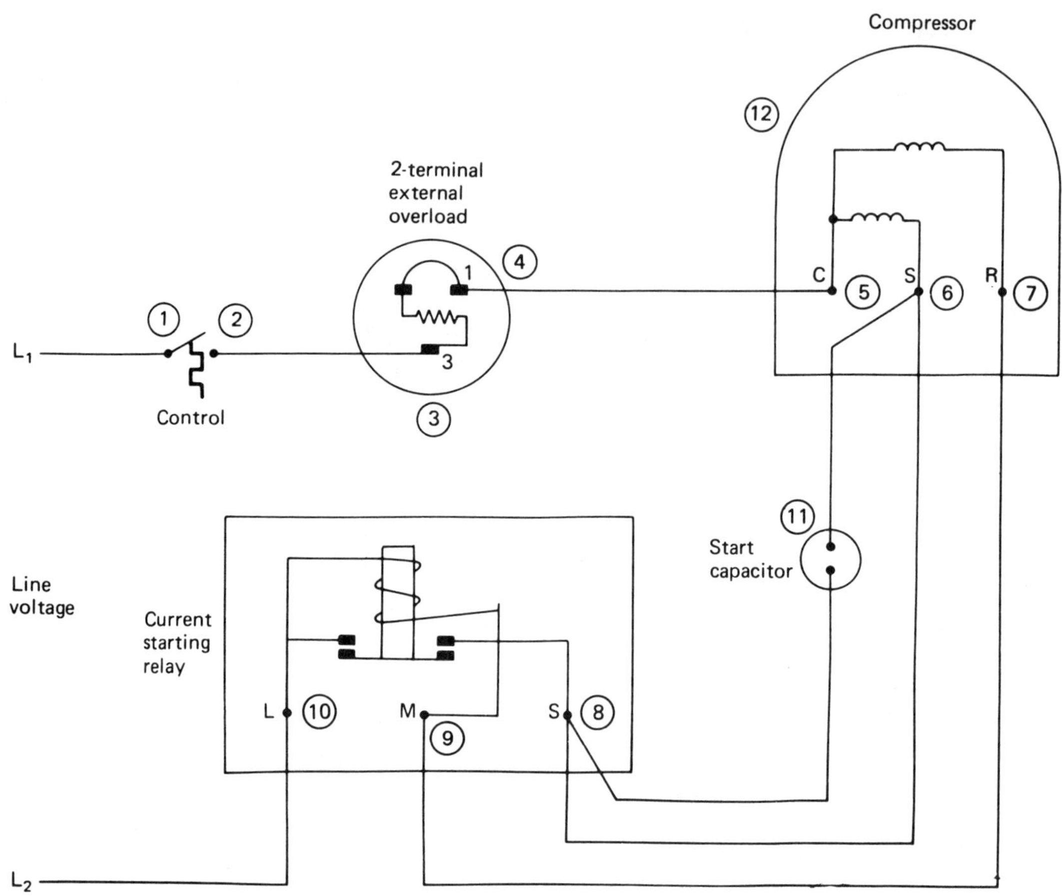

Figure 4–25 Current-type starting relay system with a two-terminal overload.

continuity is found, continue to make the following checks. If no or improper continuity is found, replace the compressor. The start winding is defective.

9. Remove the wire from point 7 (run terminal). Check the continuity between points 5 and 7. Compare this reading to that given by the manufacturer for that particular compressor motor. If the proper continuity is found, continue to make the following checks. If no or improper continuity is found, replace the compressor. The run winding is defective.
10. Check the continuity between point 5 (common terminal) and point 12 (compressor housing). If no continuity is found, continue to make

the following checks. If continuity is found, replace the compressor. The motor winding is defective.

11. Check the continuity between points 10 and 8. If no continuity is found, continue to make the following checks. If continuity is found, replace the start relay. The contacts are closed and should be open. Be sure to use an exact replacement.
12. Check the continuity between points 10 and 9. If continuity is found, continue to make the following checks. If no continuity is found, replace the relay. The coil is open. Be sure to use an exact replacement.
13. Check the continuity between point 7 (run terminal) and point 9. If continuity is found, place the wire back on the terminal and continue to make the following checks. If no continuity is found, repair the wire or loose connections.
14. Check the continuity between point 6 (start terminal) and point 8. If continuity is found, continue to make the following checks. If no continuity is found, repair the wire or loose connections.
15. If a starting capacitor is used, disconnect the wire from point 11. Set the ohmmeter on R × 100,000. Check the continuity between points 8 and 11. If needle deflection is found, continue to make the following checks. If no needle deflection is found, the capacitor is open. Replace the capacitor. Be sure to use a proper replacement.
16. Set the ohmmeter on R × 1. Check the continuity between points 8 and 11. If no continuity is found, continue to make the following checks. If continuity is found, the capacitor is shorted. Replace the capacitor. Be sure to use a proper replacement.
17. If the above steps do not locate the trouble and the compressor hums but does not start, disconnect the wire from point 8 and touch it to point 10. Turn on the electricity. If the compressor starts, immediately remove the wire from point 10. A slight spark may occur. If the compressor continues to operate, replace the relay. Be sure to use an exact replacement. If the compressor does not start, reconnect the wire to point 8 and proceed to next step.
18. Check the amperage drawn in the wire to point 5 (common terminal) while the compressor is trying to run. Compare the amperage draw to the locked rotor (LR) amperage of the motor. If the amperage drawn is very close to the LR listed by the manufacturer, replace the compressor. The compressor has mechanical problems.

Permanent Split Capacitor (PSC) System with Two-terminal Overload

PSC motors are low starting torque motors that have no starting relay or starting capacitor. This system is used on builders' models and room-type air conditioning units. Use the following procedure to check this type of system. Refer to Figure 4-26.

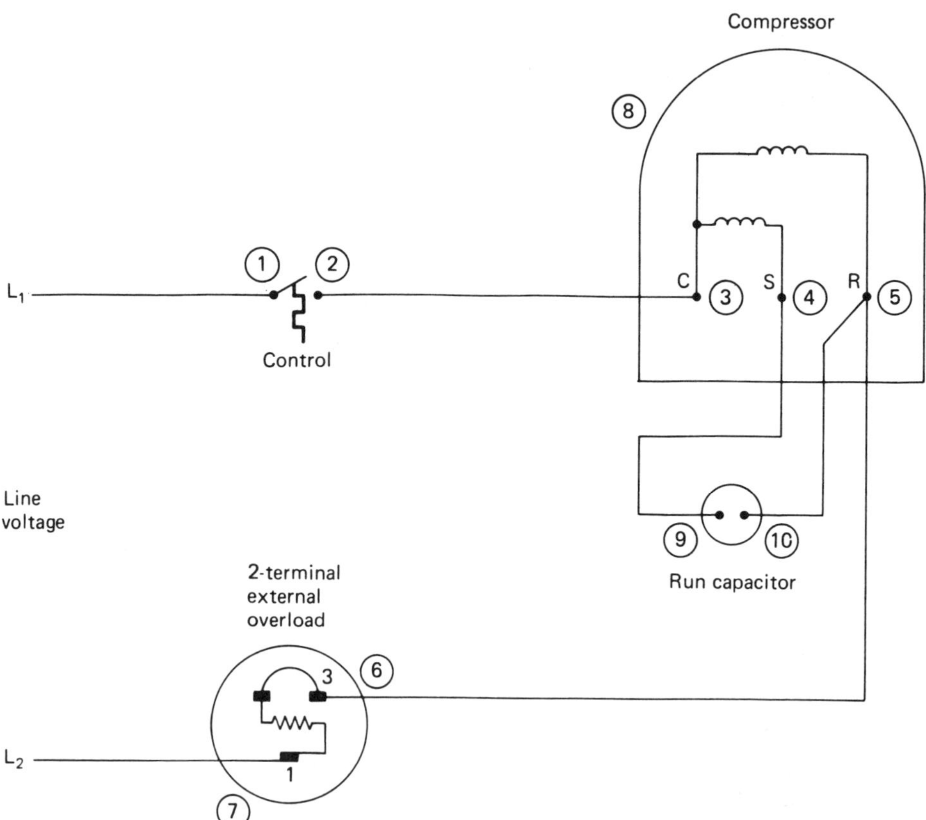

Figure 4–26 Permanent split capacitor (PSC) system with two-terminal overload.

1. Check to make certain that the proper voltage is being supplied to the unit. If proper voltage is being supplied, continue to make the following checks. If the wrong voltage is being supplied, correct the problem with the power supply.
2. Disconnect the electrical power from the unit.
3. If a fan motor is being used, disconnect one of the leads.

4. Make the following checks with an ohmmeter. Be sure to zero the ohmmeter.
5. Check the continuity between points 1 and 2. Refer to Figure 4-26. If continuity is found, continue to make the following checks. No continuity indicates the control contacts are open. Close the contacts by setting the control demand operation. If no continuity is found with the control on demand, replace the control.
5. Check the continuity between points 2 and 3. If continuity is found, continue to make the following checks. If no continuity is found, repair the wiring or loose connections.
7. Remove the wire from point 4 (start terminal). Check the continuity between points 3 and 4. Compare this reading to that given by the manufacturer for that particular compressor motor. If the proper continuity is found, continue to make the following checks. If no or improper continuity is found, replace the compressor. The start winding is defective.
8. Remove the wiring from point 5 (run terminal). Check the continuity between points 3 and 5. Compare this reading to that given by the manufacturer for that particular compressor motor. If proper continuity is found, continue to make the following checks. If no or improper continuity is found, replace the compressor. The run winding is defective.
9. Check the continuity between point 3 (common terminal) and point 8 (compressor housing). If no continuity is found, continue to make the following checks. If continuity is found, replace the compressor. The motor winding is defective.
10. Set the ohmmeter on R × 100,000. Check the continuity between points 9 and 10. If needle deflection is found, continue to make the following checks. If no needle deflection is found, the capacitor is open. Replace the capacitor. Be sure to use a proper replacement.
11. Set the ohmmeter on R × 1. Check the continuity between points 9 and 10. If no continuity is found, continue to make the following checks. If continuity is found, the capacitor is shorted. Replace the capacitor. Be sure to use a proper replacement.
12. Check the continuity between points 5 and 6. If continuity is found, continue to make the following checks. If no continuity is found, repair the wire or loose connections.
13. Check the continuity between points 6 and 7. If continuity is found, continue to make the following checks. If no continuity is found, wait

about ten minutes and check the continuity again. The overload may be tripped. If no continuity now, replace the overload. Be sure to use an exact replacement.

14. Reconnect all the wiring. Connect a 12-in. piece of electrical wire to each terminal on a 130 mfd × 370 vac start capacitor. Touch the loose end of one wire to point 4 (start terminal) and the loose end of the other wire to point 5 (run terminal). Turn on the electricity. If the compressor starts, immediately remove the two wires from points 4 and 5. If the compressor continues to run, install a hard-start kit. See Figure 4-27. If the compressor does not start or stops running when the wires are removed, proceed to the next step.

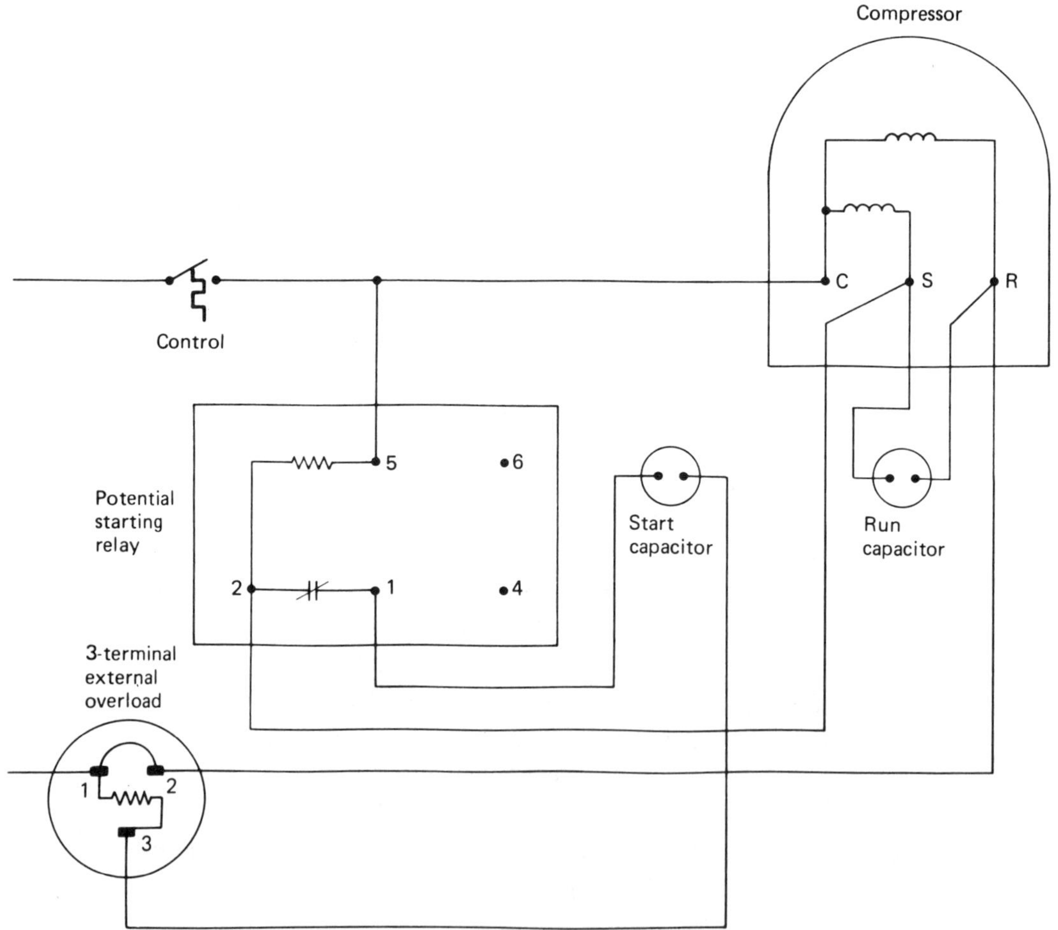

Figure 4–27 Connections used for installing a hard-start kit.

15. Check the amperage draw in the wire to point 3 (common terminal) while the compressor is trying to run. Compare the amperage draw to the locked rotor (LR) amperage of the compressor motor. If the amperage draw is very close to the LR listed by the manufacturer, replace the compressor. The compressor has mechanical problems.

Permanent Split Capacitor (PSC) System with Internal Thermostat Overload

PSC motors are low starting torque motors that have no starting capacitor or starting relay. However, they do use a run capacitor. These systems have no external overload because a more sensitive internal overload is located inside the compressor housing. This system is used on builder's models and room air conditioning units. Use the following procedure to check this type of system. Refer to Figure 4-28.

1. Check to make certain that the proper voltage is being supplied to the unit. If proper voltage is being supplied, continue to make the

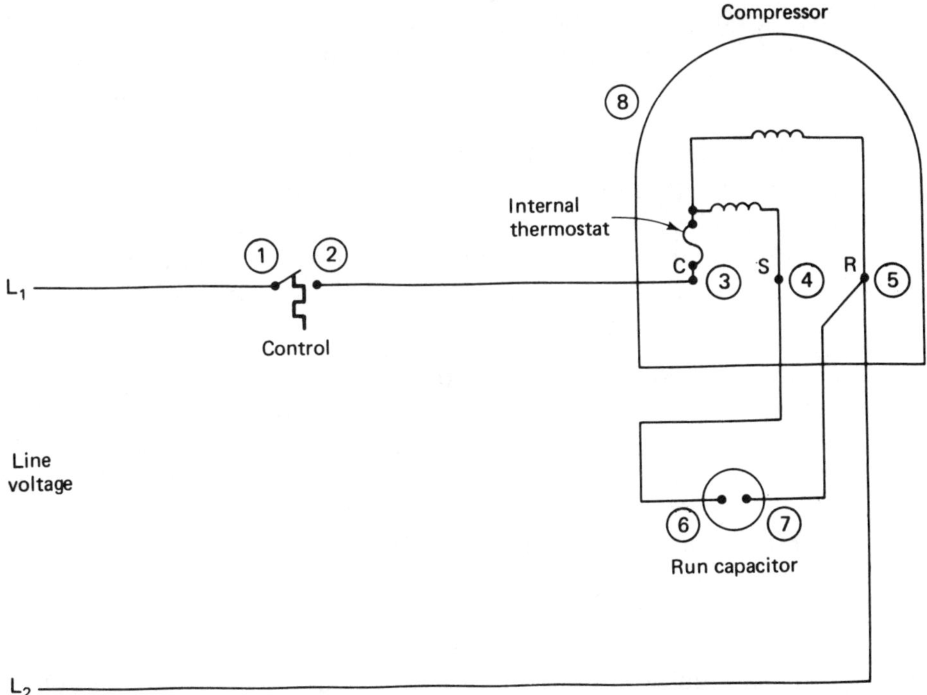

Figure 4–28 Permanent split capacitor (PSC) system with internal thermostat.

following checks. If the wrong voltage is being supplied, correct the problem with the power supply.

2. Disconnect the electrical power from the unit.
3. If a fan motor is used, disconnect one of the leads.
4. Make the following checks with an ohmmeter. Be sure to zero the ohmmeter.
5. Check the continuity between points 1 and 2. Refer to Figure 4-28. If continuity is found, continue to make the following checks. No continuity indicates the control contacts are open. Close the contacts by setting the control to demand operation. If no continuity is found with the control on demand, replace the control.
6. Check the continuity between points 2 and 3. If continuity is found, continue to make the following checks. If no continuity is found, repair the wiring or loose connections.
7. Remove the wire from point 4 (start terminal). Check the continuity between points 3 and 4. Compare this reading to that given by the manufacturer for that particular compressor motor. If continuity is found, continue to make the following checks. If no or improper continuity is found, remove the wire from point 5 (run terminal). Check the continuity between points 4 and 5. If no continuity is found, it can be assumed that the start winding is open. Replace the compressor. Check the continuity between points 3 and 5. If no continuity is found, the internal overload may be tripped. Cool the compressor shell below 125° F (51.78° C) or until the hand may be held on the compressor without much discomfort. Again check the continuity between points 3 and 4. If no continuity is found, the internal overload is defective. Replace the compressor.
8. Remove the wire from point 5 (run terminal). Check the continuity between points 3 and 5. Compare this reading to that given by the compressor manufacturer for that particular compressor motor. If proper continuity is found, continue to make the following checks. If no or improper continuity is found, check the continuity between points 4 and 5. If continuity is found, it can be assumed that the internal overload is open. Cool the compressor shell below 125° F (51.78° C) or until the hand may be held on the compressor without much discomfort. Again check the continuity between points 3 and 5. If no continuity is found, the internal overload is defective. Replace the compressor.

9. Check the continuity between points 3 and 8. If no continuity is found, continue to make the following checks. If continuity is found, replace the compressor. The motor winding is grounded.
10. Set the ohmmeter on R × 100,000. Check the continuity between points 6 and 7. If needle deflection is found, continue to make the following checks. If no needle deflection is found, the capacitor is open. Replace the capacitor. Be sure to use a proper replacement.
11. Set the ohmmeter on R × 1. Check the continuity between points 6 and 7. If no continuity is found, continue to make the following checks. If continuity is found, the capacitor is shorted. Replace the capacitor. Be sure to use a proper replacement. If the capacitor is swollen, it should be replaced.
12. Reconnect all wiring. Connect a 12-in. piece of electrical wire to each terminal on a 130 mfd × 370 vac start capacitor. Touch the loose end of one wire to point 4 (start terminal) and touch the loose end of the other wire to point 5 (run terminal). Turn on the electricity. If the compressor starts, immediately remove the two wires from points 4 and 5. If the compressor continues to run, install a hard-start kit. See Figure 4-27. If the compressor does not start or stops running when the wires are removed, proceed to the next step.
13. Check the amperage draw in the wire to point 3 (common terminal) while the compressor is trying to run. Compare the amperage draw to the locked rotor (LR) amperage of the motor. If the amperage drawn is very close to the LR listed by the manufacturer, replace the compressor. The compressor has mechanical problems.

Permanent Split Capacitor (PSC) System with Internal Thermostat Overload and External Overload

PSC motors are low starting torque motors that have no starting relay or starting capacitor. However, they do use a run capacitor. This particular compressor has both an internal thermostat and an external overload for added protection. The internal thermostat is temperature-sensitive and interrupts the line voltage to the compressor motor common terminal. The external overload is both temperature-sensitive and current-sensitive. Its contacts interrupt the control circuit. Use the following procedure to check this type of system. Refer to Figure 4-29.

1. Check to make certain that the proper voltage is being supplied to the unit. If proper voltage is being supplied, continue to make the

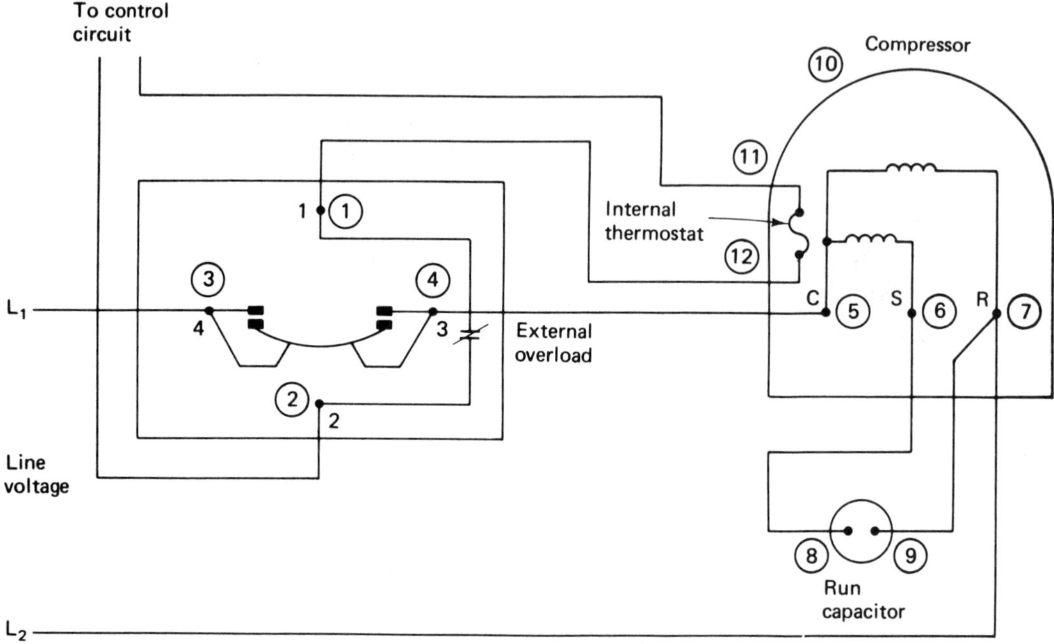

Figure 4–29 Permanent split capacitor (PSC) system with internal thermostat and external overloads.

following checks. If the wrong voltage is being supplied, correct the problem with the power supply.

2. Disconnect the electrical power from the unit.
3. If a fan motor is used, disconnect one of the leads.
4. Make the following checks with an ohmmeter. Be sure to zero the ohmmeter.
5. Check the continuity between points 1 and 2. Refer to Figure 4-29. If continuity is found, continue to make the following checks. No continuity indicates that the overload is open. Wait about ten minutes. Check the continuity again. If no continuity now, replace the external overload. Be sure to use an exact replacement.
6. Check the continuity between points 3 and 4. If continuity is found, continue to make the following checks. No continuity indicates that the external overload is defective. Replace the overload. Be sure to use an exact replacement.
7. Remove the wire from point 6 (start terminal). Check the continuity between points 5 and 6. Compare this reading to that given by the

manufacturer for that particular compressor motor. If proper continuity is found, continue to make the following checks. If no or improper continuity is found, the start winding is defective. Replace the compressor.

8. Remove the wire from point 7 (run terminal). Check the continuity between points 5 and 7. Compare this reading to that given by the manufacturer for that particular compressor motor. If proper continuity is found, continue to make the following checks. If no or improper continuity is found, the run winding is defective. Replace the compressor.
9. Remove the wire from point 11. Check the continuity between points 11 and 12. If continuity is found, continue to make the following checks. If no continuity is found, the internal thermostat is open. Cool the compressor below 125° F (51.78° C) or until the hand may be held on the housing without much discomfort. Again check the continuity between points 11 and 12. If no continuity is found, replace the compressor. The internal thermostat is defective.
10. Check the continuity between points 5 and 10. If no continuity is found, continue to make the following checks. If continuity is found, replace the compressor. The motor winding is grounded.
11. Set the ohmmeter on R × 100,000. Check the continuity between points 8 and 9. If no needle deflection is found, the capacitor is open. Replace the capacitor. Be sure to use a proper replacement.
12. Set the ohmmeter on R × 1. Check the continuity between points 8 and 9. If no continuity is found, continue to make the following checks. If continuity is found, the capacitor is shorted. Replace the capacitor. Be sure to use a proper replacement. If the capacitor is swollen, it should be replaced.
13. Reconnect all wiring. Connect a 12-in. piece of electrical wire to each terminal of a 130 mfd × 370 vac start capacitor. Touch the loose end of one wire to point 6 (start terminal) and the loose end of the other wire to point 7 (run terminal). Turn on the electricity. If the compressor starts, immediately remove the two wires from points 6 and 7. If the compressor continues to run, install a hard-start kit. See Figure 4-27. If the compressor does not start or stops running when the wires are removed, proceed to the next step.
14. Check the amperage draw in the wire to point 5 (common terminal) while the compressor is trying to run. Compare the amperage draw to the locked rotor (LR) amperage of the motor. If the amperage draw is

very close to the LR listed by the manufacturer for that particular compressor motor, replace the compressor. The compressor has mechanical problems.

4.14 PROCEDURE FOR CHECKING THERMOSTATIC EXPANSION VALVES

[This section is reprinted by permission of Alco Valve Co., St. Louis, MO.]

In checking complaints, if the expansion valve is suspected as the source of trouble, an orderly procedure for locating the exact difficulty and remedying it in the shortest time possible is desirable.

1. Check both the suction pressure and discharge pressure. They act as the "pulse" of the refrigerating system and are the best guide to locating trouble.
2. Check the superheat. For this purpose, an accurate pocket thermometer may be taped to the suction line at the remote bulb location.

The following headings group suggested possibilities of trouble indicated by the gauge and superheat readings.

Low-Suction Pressure—High Superheat

A. Expansion valve limiting flow.
   1. Inlet pressure too low from excessive vertical lift, too small liquid line, or excessively low condensing temperature. Resulting pressure differences across valve too small.
   2. Gas in liquid line—due to pressure drop in the line or insufficient refrigerant charge. If there is no sight glass in the liquid line, a characteristic whistling noise may be heard at the expansion valve.
   3. Valve restricted by pressure drop through coil, requiring change to external equalizer. See Figure 4-30.
   4. External equalizer line plugged or external equalizer connection capped without providing a new valve cage or body with internal equalizer.
   5. Moisture, wax, oil, or dirt plugging the valve orifice. Ice formation or wax at the valve seat may be indicated by a sudden rise in suction pressure after shutdown and system has warmed up.

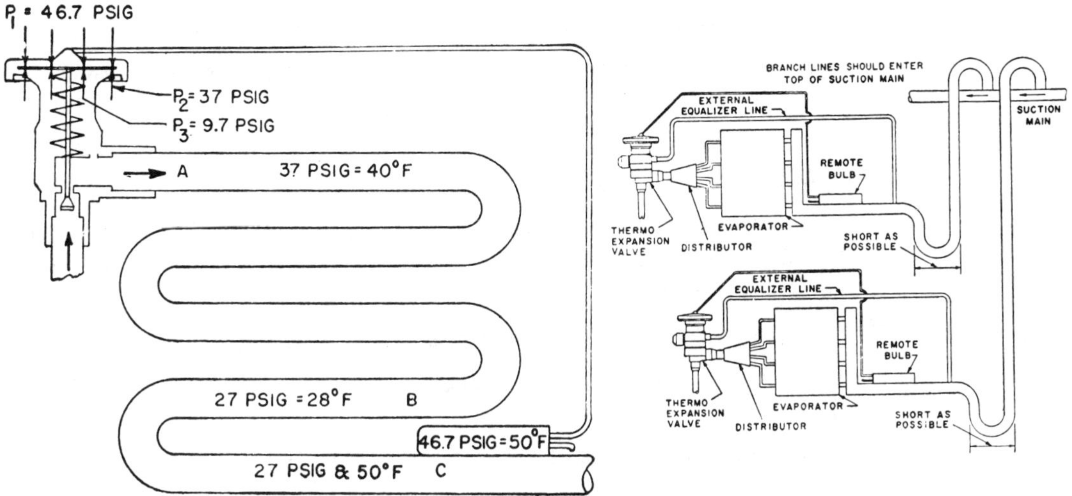

Figure 4–30 External equalizer connection methods: (a) thermo-expansion valve with internal equalizer on evaporator with 10 psi pressure drop; (b) Schematic of recommended piping of rising suction lines to a common suction main.

6. Valve orifice too small.
7. Superheat adjustment too high.
8. Power assembly failure or partial loss of charge.
9. Remote bulb of gas charged thermo expansion valve has lost control due to the remote bulb tubing or power head being colder than the remote bulb.
10. Filter screen clogged.
11. Wrong type oil.

B. Restriction in system other than thermostatic expansion valve. (Usually, but not necessarily, indicated by frost or lower than normal temperature at point of restriction.)

1. Strainers clogged or too small.
2. Solenoid valve failure or valve undersized.
3. King valve at liquid receiver outlet too small or not fully opened.
4. Plugged lines.
5. Hand valve stem failure or valve too small or not fully opened.
6. Liquid line too small.
7. Suction line too small.

8. Wrong type oil in system, blocking liquid flow.
9. Discharge or suction service valve on compressor restricted or not fully open.

Low Suction Pressure—Low Superheat

A. Poor distribution in evaporator, causing liquid to short circuit through favored passes, throttling valve before all passes receive sufficient refrigerant.

B. Compressor oversized or running too fast due to wrong pulley.

C. Uneven or inadequate coil loading, poor air distribution or brine flow.

D. Evaporator too small—indicated sometimes by excessive ice formation.

E. Evaporator oil logged.

High Suction Pressure—High Superheat

A. Compressor undersized.

B. Evaporator too large.

C. Unbalanced system having an oversized evaporator, an undersized compressor and a high load on the evaporator.

D. Compressor discharge valves leaking.

High Suction Pressure—Low Superheat

A. Compressor undersized.

B. Valve superheat setting too low.

C. Gas in liquid line with oversized thermostatic expansion valve.

D. Compressor discharge valves leaking.

E. Pin and seat of expansion valve wire drawn, eroded, or held open by foreign material resulting in liquid flood-back.

F. Ruptured diaphragm or bellows in a constant pressure (automatic) expansion valve resulting in liquid flood-back.

G. External equalizer line plugged or external equalizer connection capped without providing a new valve cage or body with internal equalizer.

H. Moisture freezing valve in open position.

Fluctuating Suction Pressure

A. Poor control.

1. Improper superheat adjustment.
2. Trapped suction line. See Figure 4-31.

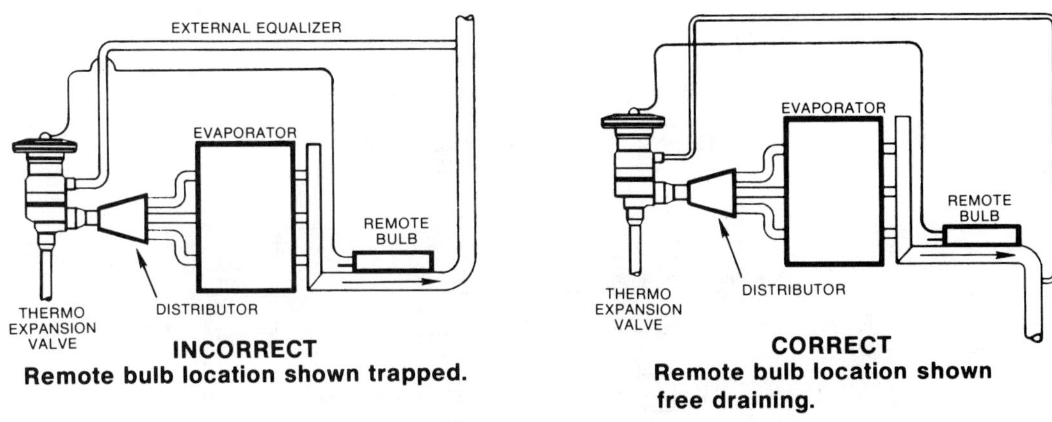

Figure 4–31 Remote bulb installed on trapped suction line.

3. Improper remote bulb location or application. See Figure 4-32.
4. "Flood-back" of liquid refrigerant caused by poorly designed liquid distribution device or uneven coil circuit loading. Also improperly hung evaporator. Evaporator not plumb. See Figure 4-33.
5. External equalizer tapped at common point on application with more than one valve on same evaporator.
6. Faulty condensing water regulator, causing change in pressure drop across valve.
7. Evaporative condenser cycling, causing radical change in pressure difference across expansion valve.
8. Cycling of blowers or brine pumps.

Fluctuating Discharge Pressure

A. Faulty condensing water regulating valve.

B. Insufficient charge—usually accompanied by corresponding fluctuation in suction pressure.

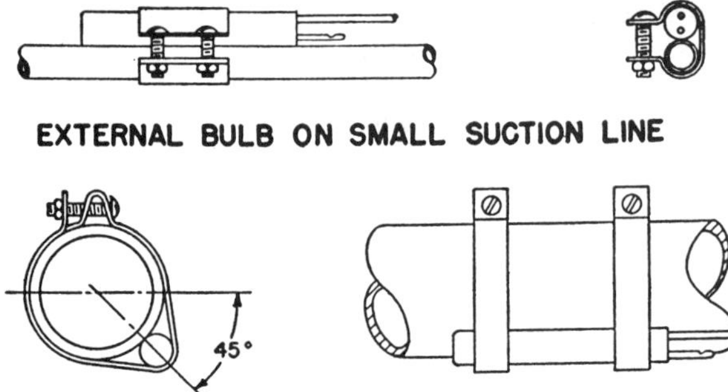

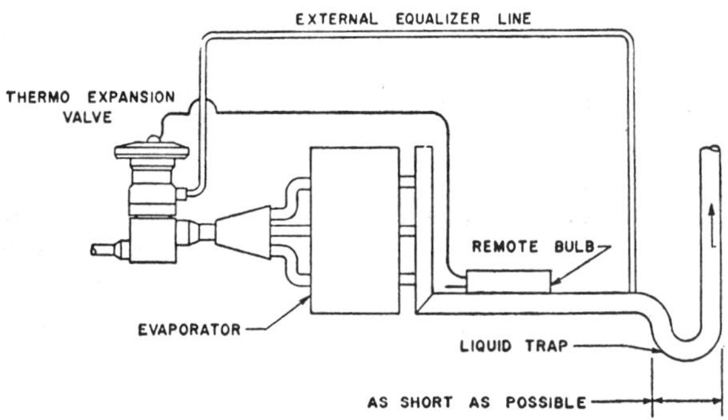

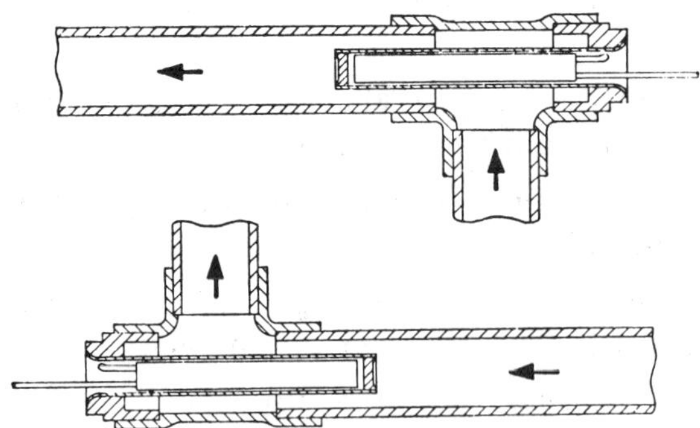

Figure 4–32 Recommended remote bulb location and schematic piping for rising suction line.

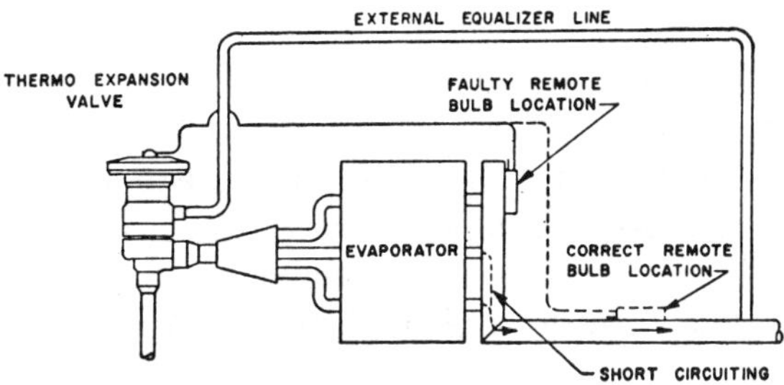

Figure 4–33 Correct remote bulb location on short circuiting evaporator to prevent "flood-back."

C. Cycling of evaporative condenser.

D. Inadequate and fluctuating supply of cooling water to condenser.

High Discharge Pressure

A. Insufficient cooling water (Inadequate supply or faulty water valve).

B. Condenser or liquid receiver too small.

C. Cooling water above design temperature.

D. Air or noncondensable gases in condenser.

E. Overcharge of refrigerant.

F. Condenser dirty.

G. Air type condenser improperly located to dispel hot discharge air.

Expansion Valve "Freeze-Ups"

Expansion valve trouble may occur from the formation of ice crystals or the separation of wax out of the oil at the pin seat, causing the refrigerant flow to be restricted or stopped entirely. Waxing generally occurs only at extreme low temperatures; however, ice formation due to moisture in the system can occur whenever the evaporator is operating below freezing. Moisture is generally indicated by the starving of the evaporator. However, it may cause a valve to freeze open and cuase "flood-back." When the compressor stops and the expansion valve and the system are allowed to warm up, the system will operate satisfactorily until the ice crystals form again. Another indication of moisture in the expansion valve can be noted if tapping the valve body will cause it to feed again for a short time. Under no circumstances should a torch be used on the valve body to melt ice accumulation at the pin and seat. If the ice must be melted to put the system in operation temporarily, then apply hot rags. This is equally effective and will not damage the valve.

To remedy this, install a drier of suitable size and operate the system above freezing for a few days to allow the moisture to be trapped in the drier rather than to freeze at some location in the evaporator. If moisture persists, then the refrigerant charge and the oil should be dumped, and the system dried by means of a vacuum and heat or by blowing dry nitrogen gas through the system.

Moisture can be admitted to the system:

A. At the time of original installation from moist air in the piping.
B. At the time of charing by not blowing the air and moisture out of the charging hose before tightening the fitting at the connection.

4.15 PROCEDURE FOR TORCH BRAZING

[Article by Englehard Industries, Warwick, RI. Used with permission.]

The article describes procedures that are important for making sound brazements of tubing in the air conditioning and refrigeration industry, using the phos-copper and the silver brazing filler metals.

The importance of brazing cannot be overlooked and/or overemphasized since it is a major part of the air conditioning and refrigeration industry.

Brazing is the process used in joining the major components of a refrigeration system into a closed circuit. Since the closed circuit contains refrigerant, every brazed joint must be leak-free; if not, the refrigerant will escape, creating a severe inconvenience to the customer as well as a costly repair.

The purpose of this article is to provide the basic information about the correct method for torch brazing.

Definitions

Brazing is the application of heat above a temperature of 800° F and below the melting points of the base metals to produce coalescence or bonding by surface adhesion forces between the molten filler metals and the surfaces of the base metals. The filler metal is distributed through the joint by using capillary action.

Brazing is not to be confused with soldering, even though the procedures are very similar. Soldering is a term used for metal-joining processes at temperature below 800° F.

In order for the bonding and distribution of capillary attraction to occur, the filler metal must be able to "wet" the base metals. Wetting is the phenomenon in which the forces of attraction between the molecules of the molten filler metal and the molecules of the base metals are greater than the inward forces of attraction existing between the molecules of the filler metal.

The degree of wetting is a function of the compositions of the base metal and the filler metal and the temperature. Good wetting can only occur on perfectly clean and oxide-free surfaces.

Filler Metals

The quality and strength of a brazement are more a function of the physical parameters of the joint and the brazing procedures used than which filler metal is applied to the joint. These parameters determine the selection of the best-suited and the most easily handled filler metal for a particular joint.

The phos-copper brazing alloys were specifically developed for joining copper and copper alloys. They are used for brazing copper, brass, bronze, or combinations of these.

When brazing with either brass or bronze, a flux must be used to prevent the formation of an "oxide coat" over these base metals. This coat would prevent wetting and the flow of the filler metal. However, when brazing copper to copper joints, these alloys are self-fluxing.

Because of the phosphorous embrittlement of the joint, phos-copper filler metals should not be used on ferrous metals or base metals containing more than 10 percent nickel. These alloys are also not recommended for use on aluminum bronze.

Unlike the phos-copper alloys, the silver brazing alloys contain no phosphorous. These filler metals are used to braze all ferrous, copper, and copper-based metals with the exception of aluminum and magnesium. They require a flux with all applications.

Torch Brazing Procedures

Copper to copper joints using the phos-copper brazing alloys

I. Correct oxygen and fuel gas mixture. See Figure 4-34.

Excessive gas mixture—a reducing-type flame denotes excessive fuel gas; a greater amount of fuel gas than oxygen. A slightly reducing flame heats and cleans the metal surface for quicker and better brazing. See Figure 4-35.

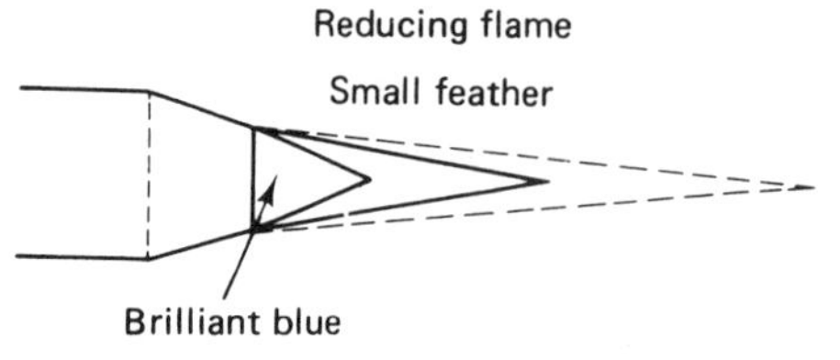

Figure 4–34 Best type of flame for brazing.

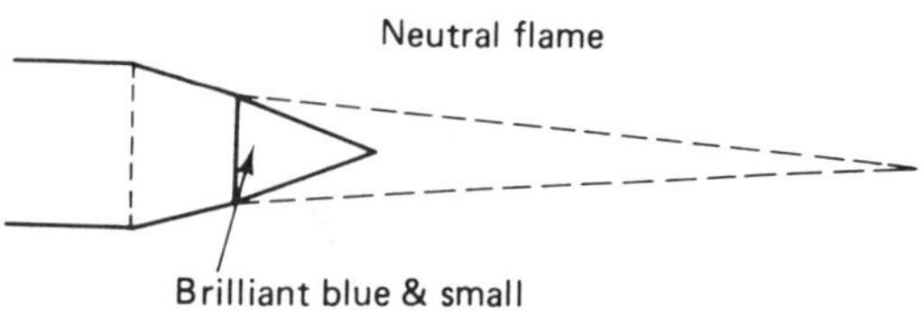

Figure 4–35 Feather or reducing flame disappeared.

Balanced gas mixture—the gas mixture contains an equal amount of oxygen and fuel gas. It produces a flame that heats the metal, but has no other effect. See Figure 4-36.

Excessive oxygen mixture—the gas mixture contains an excessive amount of oxygen. It produces a flame that oxidizes the metal surface. A black oxide scale will form on the metal surface.

II. Cleanliness. General cleanliness is of prime importance to reliable brazing. All metal surfaces to be brazed must be cleaned of all dirt and foreign matter. When repair work is to be done, the metal surfaces to be joined must be either wire-brushed and/or cleaned with sand paper. A concerted effort must be made to keep oil, paint, dirt, grease, and aluminum off the surface of metals to be joined for these contaminants will:

1. Keep brazing filler metal from flowing into the joint.
2. Prevent brazing filler metal from wetting or bonding to the metal surfaces.

III. Correct insertion and clearance of parts. Minimum insertion distance is equal to the diameter of the inner tube. There should be 0.001–0.005 in. clearance between the walls of the inner and outer tubes. See Figure 4-37.

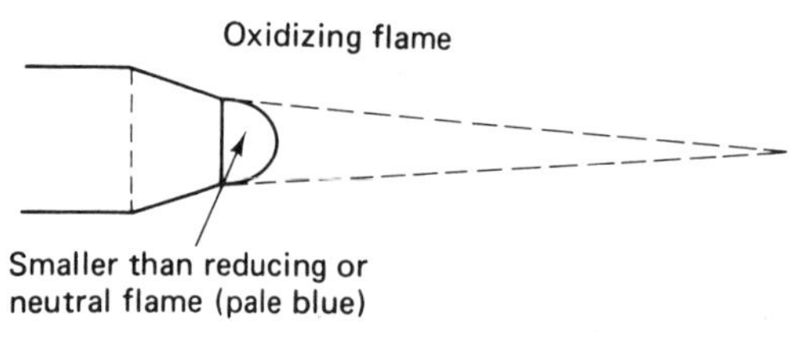

Figure 4–36 Worst type of brazing flame.

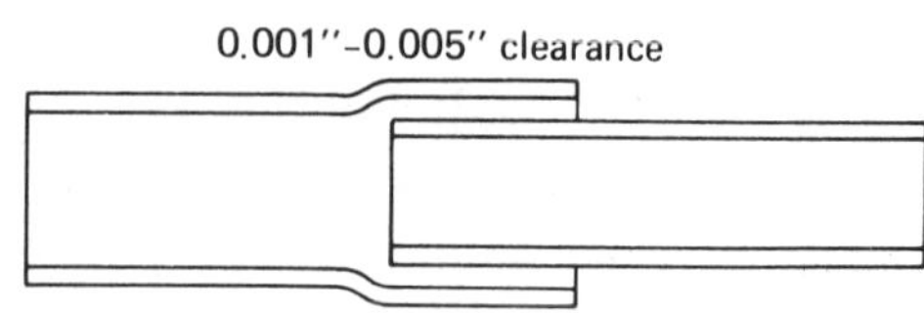

Figure 4–37 Correct clearance on joint parts.

IV. Correct amount of application of heat. Heat evenly distributed over the entire joint circumference and joint length. Use enough heat to get the entire joint hot enough to melt the filler metal without heating the filler metal directly with the flame. See Figure 4-38.

Heat both tubes at the joint, and distribute heat evenly. Do not overheat the joint to the point that the metal to be joined begins to melt. See Figure 4-39.

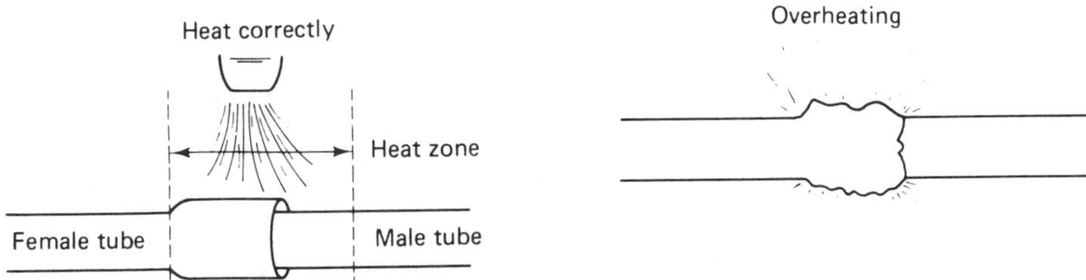

Figure 4–38 Proper heat distribution over joint.

Figure 4–39 Overheated joint.

Always try to use the correct size torch tip, and a slightly reducing flame. Overheating enhances the base metal-filler metal interactions (e.g., formation of chemical compounds). In the long run, these interactions are detrimental to the life of the joint. Some examples of heating methods are shown in Figures 4-40 through 4-42.

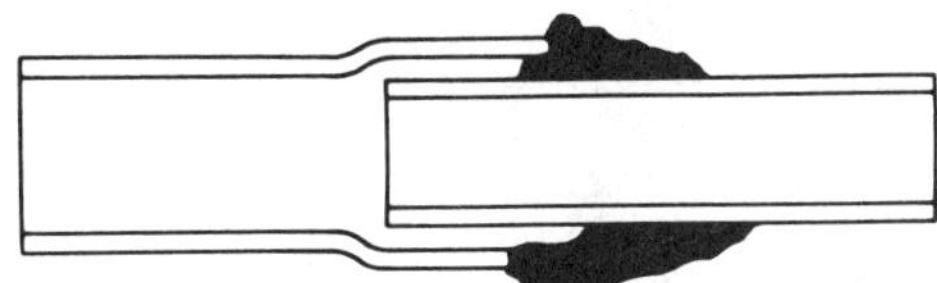

Figure 4–40 Female tube too cool.

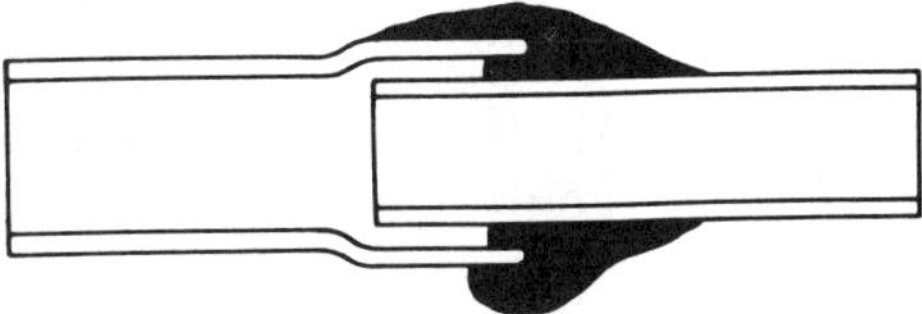

Figure 4–41 Filler metal and torch applied at same time.

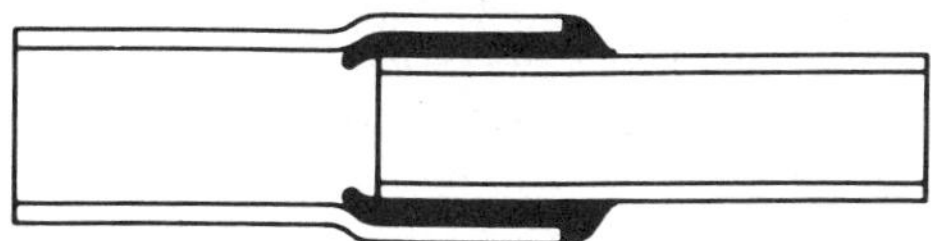

Figure 4–42 Properly made joint.

V. Correct application of rod. See Figure 4-43.

Use the brazing rod as a temperature indicator. If the brazing rod melts when placed on the hot joint, the tubing is hot enough to begin brazing. For best results, preheat the rod slightly with the outer envelope of the flame, but the hot copper (not the direct flame) should melt the brazing rod.

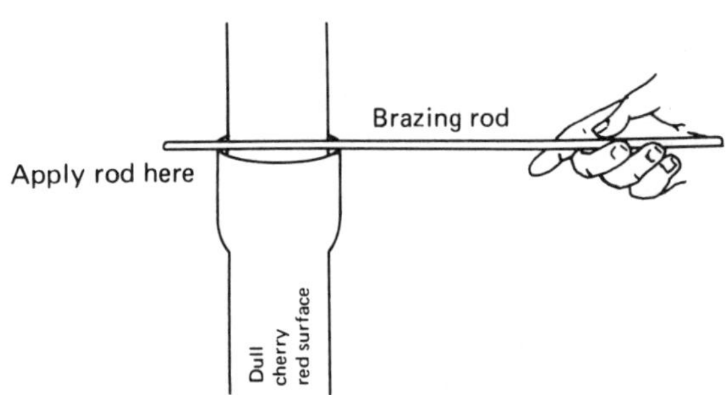

Figure 4–43 Correct application of brazing rod.

VI. Capillary attraction. See Figure 4-44. This is the phenomenon by which the brazing filler metal is drawn into the joint. It is caused by the attraction between the molecules of the brazing filler metal and the molecules of the metal surfaces to be joined. It can only work if: (1) the surface of the metal is clean; (2) the clearance between the metal surfaces is correct; (3) the metal at the joint area is hot enough to melt the brazing filler metal; and (4) the brazing filler metals will flow toward the heat source as illustrated above.

Copper to brass tube joints using the phos-copper brazing alloys

I. The above-mentioned procedures for copper joints are strictly followed.

II. Before application of heat to the joint, a small amount of flux is applied to allow wetting of the brass by the filler metal.

III. Upon completion of the brazement, the flux residue is thoroughly cleaned off. Cleaning can be accomplished with hot water and mechanical brushing. Most fluxes are corrosive and must be completely removed from the joint.

Steel to steel, copper, brass, or bronze using silver brazing alloys.

I. The above-mentioned procedures for joining copper to copper are strictly followed.

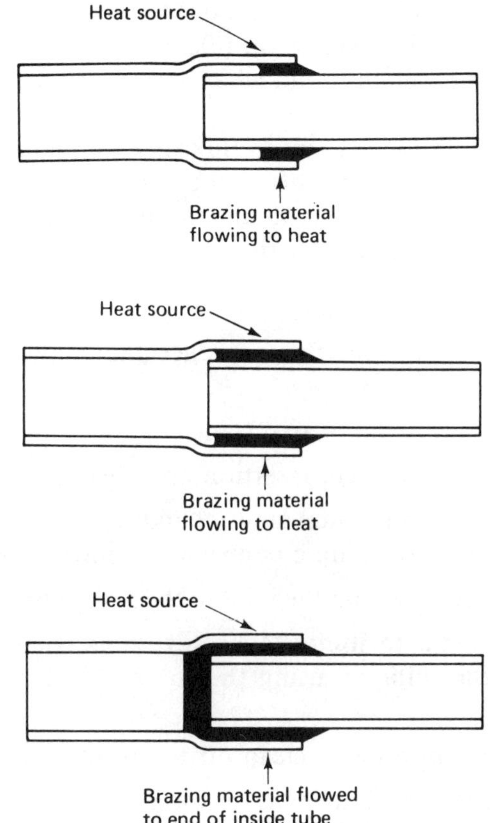

Figure 4–44 The capillary attaction process.

II. Before application of heat to the joint, flux is applied to allow wetting and flow to take place between the filler metal and the base metals.

III. Before application of the filler metal to the joint, a small amount of flux is applied to the rod. Rod is heated up and then dipped into the flux. This coats the filler metal with a thin layer of flux that provides for quick flow by preventing the formation of an oxide coat around the filler metal (zinc oxide).

IV. Flux is completely washed off upon completion of the joint.

Fluxes

A flux is analogous to a sponge absorbing water, only a flux is absorbing oxides. It can only absorb so much before it becomes useless as a flux.

When a flux becomes saturated with oxides, its viscosity increases. The ability of the flux to be displaced by the filler metal is therfore impeded. This leads to flux intrapment in the joint, which eventually causes corrosion and leaks.

One should use the least amount of flux that will get the job done. Then thoroughly clean the flux residue off after completing the brazement. Apply the flux along the surface of the joint and not into the joint itself. Allow the flux to flow into the joint head of the filler metal.

Simplified Rules For Good Brazing

1. Use a slightly reducing flame to get the maximum heating and cleansing action.
2. Make sure metal surfaces are clean.
3. Examine parts for correct insertion and clearances.
4. Flux parts: (1) use minimum amount and (2) apply to outside of joint. No flux required for joining copper with phos-copper filler metals.
5. Apply heat evenly around the joint to the proper brazing temperature.
6. Apply filler metal to joint. Make sure it is completely distributed throughout the joint by using the torch. Molten filler metals will follow the heat.
7. If flux is used, thoroughly clean off the residue.
8. A major part of good brazing is to complete the brazement as quickly as possible. Keep the heating cycle short. Avoid overheating.
9. Have adequate ventilation. Brazing may result in hazardous fumes: cadmium fumes from cadmium-bearing filler metals and fluoride fumes from fluxes.

CHAPTER 5

Electrical Troubleshooting

5.1 IF THE COMPRESSOR WILL NOT RUN

If a motor compressor fails to start and run properly, it is important that it be tested to determine its condition. It is possible that external electrical components may be defective, the protector may be open, a safety device may be tripped, or other conditions may be preventing compressor operation. If the motor compressor is not the source of the malfunction, replacing the compressor will only result in the unnecessary expenditure of time and money, while the basic problem remains.

1. If there is no voltage at the compressor terminals, follow the wiring diagram and check back from the compressor to the power supply to find where the circuit is interrupted.

   Check the controls to see if the contact points are closed (low pressure control, high pressure control, thermostat, oil pressure safety control, etc.). If a contactor is used, check to see if the contacts are closed. Check for a blown fuse, open disconnect switch, or loose connection.

2. If power is available at the compressor terminals, and the compressor does not run, check the voltage at the compressor terminals while attempting to start the compressor.

   If the voltage at the compressor terminals is below 90 percent of the nameplate voltage, it is possible that the motor may not develop sufficient torque to start. Check to determine if wire sizes are adequate, electrical connections are loose, the circuit overloaded, or if the power supply is adequate.

3. On units with single-phase PSC motors, the suction and discharge pressures must be equalized before starting because they are low starting torque motors. Any change in the refrigerant metering device, the addition of a drier, or other changes in the system components may delay pressure equalization and create starting difficulties. If PSC motor starting problems are being encountered, the addition of a hard start kit is recommended.

4. On single-phase compressors, a defective capacitor or starting relay may prevent the compressor starting. If the compressor attempts to start but is unable to do so, or if there is a humming sound, check the starting relay to see if the relay contacts are damaged or fused. The relay points should be closed during the initial starting cycle, but should open as soon as the compressor comes up to speed.

   Remove the wires from the starting relay and capacitors. Use a high voltage ohmmeter to check for continuity through the relay coil. Replace the relay if there is no continuity. Use an ohmmeter to check across the relay contacts. Potential relay contacts are normally closed when the relay is not energized; current relay contacts are normally open. If either gives an incorrect reading, replace the relay.

   Any capacitor found to be bulging, leaking, or damaged should be replaced.

   Make sure capacitors are discharged before checking. Check for continuity between each capacitor terminal and the case. Continuity indicates a short, and the capacitor should be replaced.

   Substitute a "known to be good" start capacitor if available. If the compressor then starts and runs properly, replace the capacitor. On PSC motors, substitute "a known to be good" run capacitor if available. If the compressor then starts and runs properly, replace the original run capacitor.

   If a capacitor is not available, an ohmmeter may be used to check run and start capacitors for short or open circuits. Use an ohmmeter set to its highest resistance scale, and connect the prods to the capacitor terminals.

   a. With a good capacitor, the indicator should first move to zero, and then gradually increase to infinity.

   b. If there is no movement of the ohmmeter needle, an open circuit is indicated.

   c. If the ohmmeter needle moves to zero and remains there or on a low resistance reading, a short circuit is indicated.

   Defective capacitors should be replaced.

5. If the correct voltage is available at the compressor terminals, and no current is drawn, remove all wires from the terminals and check for continuity through the motor windings. On single-phase motor compressors, check for continuity from terminal C to R, and C to S. On three-phase compressors, check for continuity between the terminals for connections to phases 1 and 2, 2 and 3, and 1 and 3. On compressors with line break inherent protectors, an open overload

protector can cause a lack of continuity. If the compressor is warm, wait one hour for the compressor to cool and recheck. If continuity cannot be established through all motor windings, the compressor should be replaced.

Check the motor for a ground by means of a continuity check between the common terminal and the compressor shell. If there is a ground, replace the compressor.

6. If the compressor has an external protector, check for continuity through the protector or protectors.

   All external and internal inherent protectors on Coplematic compressors can be replaced in the field. On larger compressors with thermostats, thermotectors, or solid state sensors in the motor windings (D, H, S protection), the internal protective devices cannot be replaced and the stator or compressor must be changed if the internal protectors are defective or damaged.

5.2 IF THE COMPRESSOR STARTS BUT TRIPS REPEATEDLY ON THE OVERLOAD PROTECTOR

1. Check the compressor suction and discharge pressures while the compressor is operating. Be sure the pressures are within the limitations of the compressor. If the pressures are excessive, it may be necessary to clean the condenser, purge air from the system, and add a crankcase regulating valve, modify the system control, or take such other action as may be necessary to avoid excessive pressures.

   An extremely low suction pressure may indicate a loss of charge, and a suction-cooled motor compressor may not be getting enough refrigerant vapor across the motor for proper cooling.

   On units with no service gauge ports where pressures cannot be checked, check the condenser to be sure it is clean and the fan is running. Excessive temperatures on the suction and discharge line may also indicate abnormal operating conditions.

2. Check the line voltage at the motor terminals while the compressor is operating. The voltage should be within 10 percent of the nameplate voltage rating. If outside those limits, the voltage supply must be brought within the proper range or a motor compressor with different electrical characteristics must be used.

3. Check the amperage drawn while the compressor is operating. Under normal operating conditions, the amperage drawn will seldom exceed

100 percent of the nameplate amperage and should never exceed 120 percent of the namemplate amperage. High amperage can be caused by low voltage, high head pressure, high suction pressure, compressor mechanical damage, defective running capacitors, or a defective starting relay.

On three-phase compressors, check the amperage in each line. One or two high amperage legs on a three-phase motor indicates an unbalanced voltage supply, or a winding imbalance. If all three legs are not drawing approximately equal amperage, temporarily switch the leads to the motor to determine if the high leg stays with the line; the problem is in the line voltage supply. If the high amperage reading stays with the terminal, the problem is in the motor.

If the amperage is sufficiently unbalanced to cause a protector to trip, and the voltage supply is unbalanced, check with the power company to see if the condition can be corrected. If the voltage supply is balanced, indicating a defective motor phase, the compressor should be replaced.

4. Check for a defective running capacitor or starting relay in the same manner described in the previous section.
5. Check the wiring against the wiring diagram in the terminal box. On dual voltage motors, check the location of the terminal jumper bars to be sure that they are properly connected.
6. Overheating of the cylinders and head can be caused by a leaking valve plate. To check, close the suction service valve and pump the compressor into a vacuum. Stop the compressor and crack the suction valve to allow the pressure on the suction gauge to build to 0 psig (100.989 kPa). Again close the valve. If the pressure on the gauge continues to increase steadily, the valve plate is leaking. Remove the head and check the valve plate; replace if necessary.
7. If all operating conditions are normal, the voltage supply at the compressor terminals balanced and within limits, the compressor crankcase temperature within normal limits, and the amperage drawn within the specified range, the motor protector may be defective, and should be replaced.

   If the operating conditions are normal and the compressor is running excessively hot for no observable reason, or if the amperage drawn is above the normal range and sufficient to repeatedly trip the protector, the compressor has internal damage and should be replaced.

5.3 IF THE COMPRESSOR RUNS BUT WILL NOT REFRIGERATE

1. Check the refrigerant charge, and the operating pressures. Any abnormal operating condition must be corrected.
2. If the suction pressure is high, and the evaporator and condenser are functioning normally, check the compressor amperage draw. An amperage draw near or above the nameplate rating indicates normal compressor operation, and it is possible the compressor or unit may have damaged valves or does not have sufficient capacity for the application.

   An amperage draw considerably below the nameplate rating may indicate a broken suction reed or broken connecting rod in the compressor. Check the pistons and valve plate on an accessible hermetic compressor. If no other reason for lack of capacity can be found, replace a welded compressor.

5.4 READING A WIRING DIAGRAM

It is important that the service technician be able to read a unit wiring diagram or schematic during the troubleshooting procedure. When the technician has gained the ability to read these diagrams, the operating sequence of the unit can be determined from them. The following paragraphs present one method of reading wiring schematics. Each person will develop his or her own method that best suits their own way of troubleshooting equipment.

Cooling Operation: Electrical Components and Sequence

Step 1: Switches in "off" position (disconnect turned "on"). This position supplies line voltage to the sump heater (6) and transformer (7) for control voltage. ODS (12) is energized through ODA (J) in the transformer at all times. See Figure 5-1.

Step 2: First-stage cooling [system switch (A) turned to auto]. The first stage on the thermostat calls for cooling; note the operational changes. TSC-1 (D) closes, completing the 24-volt circuit through (0), energizing the switchover valve solenoid coil (11); nothing else occurs. (This is TSC-1's only function.)

Step 3: Second-stage cooling [after a 0.7 to 1.5°F (0.392°C to 0.84°C) temperature rise]. The thermostat second stage indicates that cooling is needed; note the operational changes. TSC-2 (F) now closes completing the 24-volt circuit through the Y terminal to the MS relay coil (14) and through terminal G to the F relay coil (13).

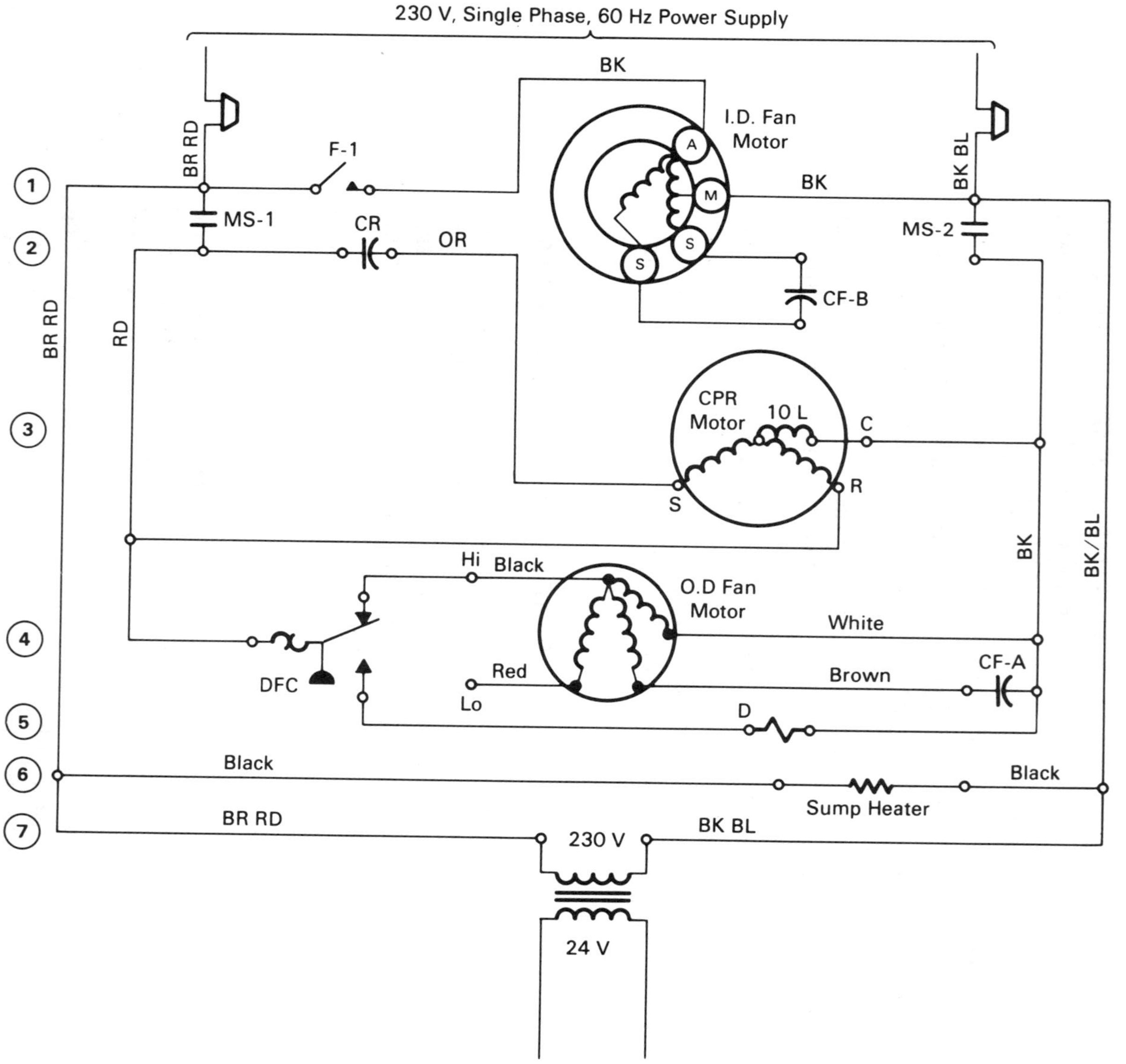
230 V, Single Phase, 60 Hz Power Supply
BK
I.D. Fan Motor
A
M
S
S
BR RD
F-1
BK
BK BL
MS-1
CR
OR
MS-2
CF-B
BR RD
RD
CPR Motor
10 L
C
S
R
BK
BK/BL
Hi
Black
O.D Fan Motor
White
DFC
Red
Lo
Brown
CF-A
D
Black
Sump Heater
Black
BR RD
230 V
BK BL
24 V
1
2
3
4
5
6
7

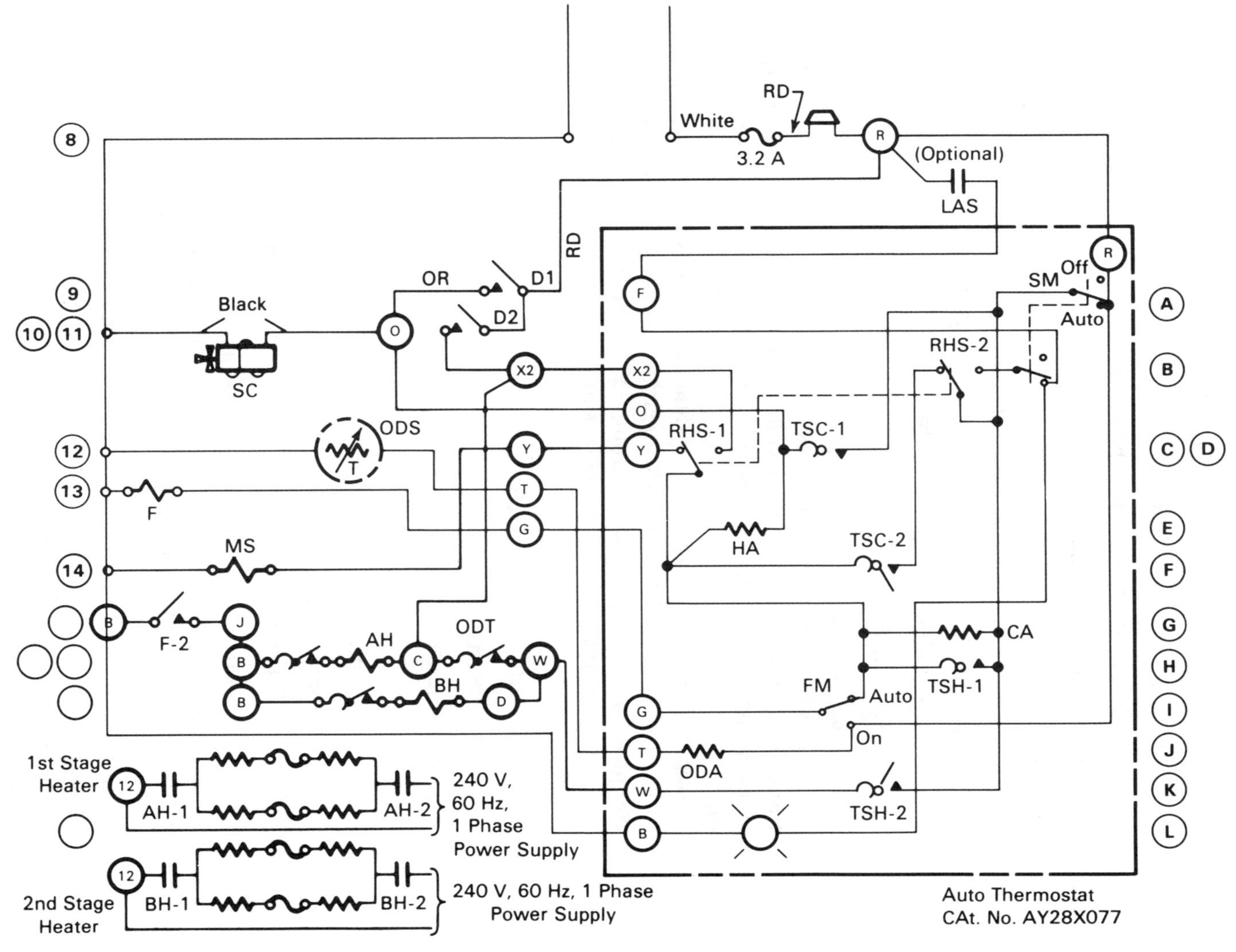

Figure 5–1 Typical Weathertron heat pump schematic diagram, cooling. (Courtesy of the Trane, Co., Dealer Products Group).

Simultaneous occurrences:

1. The MS-1 and MS-2 contacts (2) close.
2. The F-1 contacts (1) close.
3. The F-2 contacts (15) not shown close.
4. The compressor and outdoor fan start.
5. The indoor fan starts.
6. The safties interlock in the electric heater circuit.
7. Normally, TSC-1 remains energized during the cooling season and TSC-2 controls the system operation.

Heating Operation

Step 1: System switches in "off" position (the disconnect turned "on"). See Figure 5-2. In this position voltage is supplied to the sump heater (6) and the transformer (7) for the control voltage. The ODS (12) stays energized through the ODA (J) in the thermostat whenever the disconnect is turned on.

Step 2: First-stage heating [the system switch (A) is turned to auto]. The thermostat calls for heat; note the operational changes. Switch TSH-1 (H) completes the 24-volt circuit through Y to the MS relay coil (14) and through G to the F relay coil (13), simultaneously closing the following contacts:

1. MS-1 and MS-2 contacts (2)
2. F-1 contacts (1)
3. F-2 contacts (15)

Then:

1. The compressor and outdoor fan start.
2. The indoor fan starts.
3. The safeties interlock in the electric heating circuit.

The refrigerant system is now operating in the heating mode; note that the S.O.V. is de-energized. A circuit from TSH-1 (H) to HA (E) in series with SC (11) is also completed. The fixed heat anticipator (E) prevents the SC (11) from energizing because HA = 3000 ohms and the SC = 5 ohms. This circuitry eliminated an extra low-voltage wire in the field wiring.

Step 3: For second-stage heating [after 0.7 to 1.5°F (0.392 to 0.84°C) temperature drop]. The thermostat indicates that additional heating is required; note the operational changes. Switch TSH-2 (K) closes, completing the 24-volt circuit to terminal W through ODT (17) (if closed), energizing the AH relay coil (16) and BH relay coil (18) to turn on the supplementary heat. Switch TSH-2 will control the electric heat, and TSH-1 will operate the compressor until the thermostat is satisfied.

Defrost Cycle

During normal heating cycle operation, the outdoor coil acts as an evaporator and during certain conditions [below 40°F (4.4°C)] will accumulate frost or ice and require defrosting. See Figure 5-3.

Defrost Initiation: When the defrost control, DFC, air switch senses a pressure differential of approximately 0.8 in (20.32 mm) water column, and when the temperature of the liquid line supplying refrigerant to the outdoor coil is 32°F (0°C) or below, the unit will switch into defrost. The defrost control DFC (4) switches, energizing the D relay coil (5) while simultaneously turning off the OD fan. D-1 (9) and D-2 (10) contacts close. Contact D-1 (9) completes the 24-volt circuit to the SC (11) switching the S.O.V. D-2 (10) contacts, completing a circuit through X2 to C and energizes the AH relay coil (16) provided that the F relay coil (13) is energized to bring on some electric heat to temper the ID air during the defrost cycle.

Defrost Termination: When the outdoor coil liquid line temperature reaches 55°F (12.8°C) the defrost cycle will be completed. The defrost control, DFC (4), will switch to normal operating position. The D relay coil (5) is de-energized, opening contacts D-1 (9) and D-2 (10) and breaking the 24-volt circuit to the S.O.V. and the supplementary heater control circuit.

Resistance Heat Operation

In the event of refrigerant system failure or other mechanical failure, the Weathertron thermostat contains an emergency electric heat switch to electrically isolate the Weathertron system and protect it from further damage while allowing the supplementary heat to be automatically controlled by the thermostat. To activate the emergency heat switch, remove the thermostat cover and push the switch to the right. The switch is located behind the black metal system/fan switch support bar. RHS-1 (C) and RHS-2 (B) are mechanically connected and move simultaneously when the switch is activated. A signal light is energized and will burn continuously to remind the customer that the system is operating on resistance heat.

Resistance heat switch RHS-1 (C) and RHS-2 (B) switched. RHS-1 (C) moves from the Y terminal to the X2 terminal internally within the thermostat. This electrically isolates the MS relay coil (14) and prevents it from energizing on a heat call from the thermostat. When heating is needed, switch TSH-1 (H) will close, completing the 24-volt circuit through RHS-1 (C) to X2, energizing the AH relay coil (16) and BH relay coil (18) if the ODT (17) is closed. If the ODT (17) is open, BH relay coil (18) will be energized through TSH-2 (K) when needed.

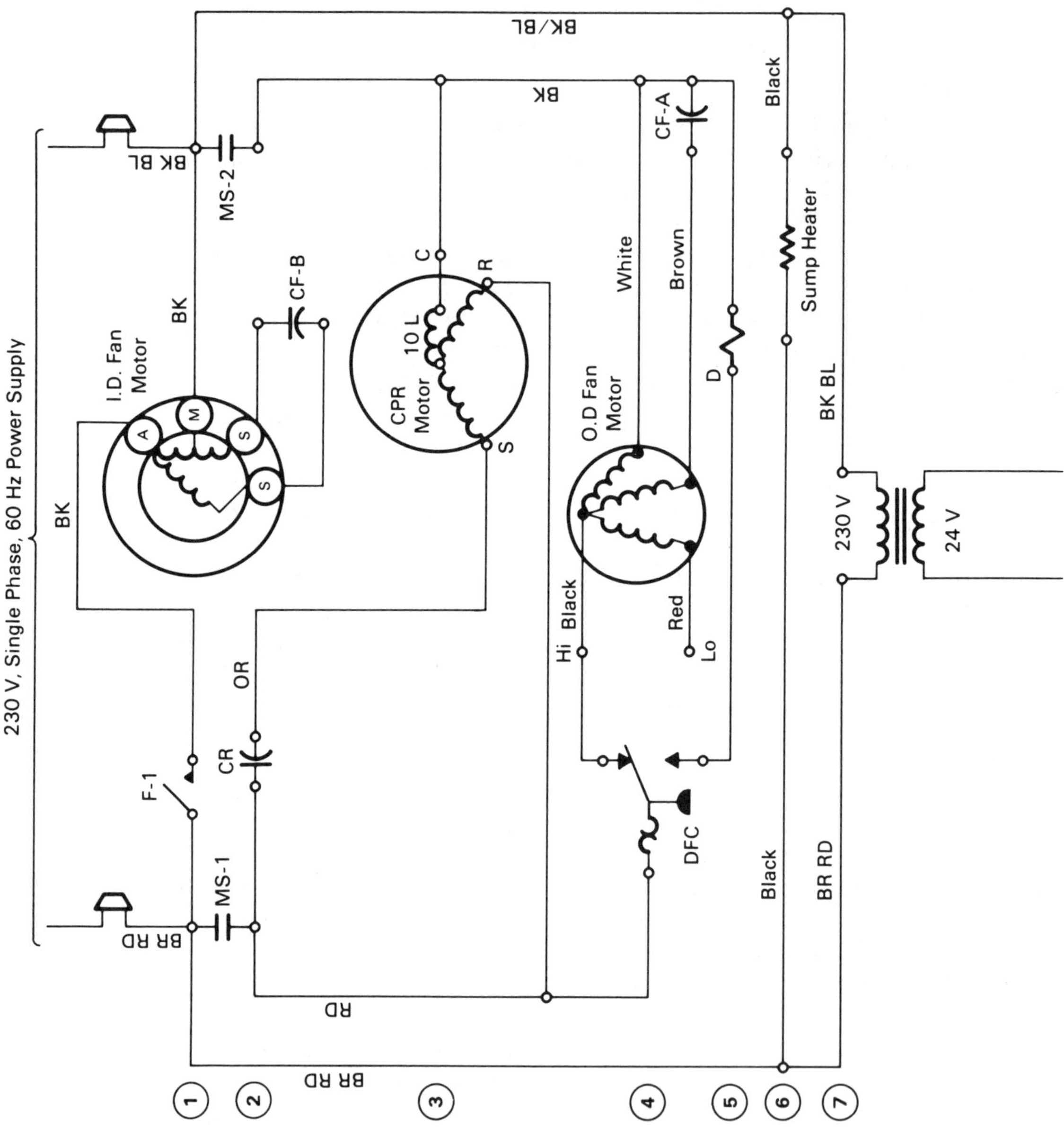
230 V, Single Phase, 60 Hz Power Supply
BK/BL
BK
BK BL
MS-2
CF-B
BK
I.D. Fan Motor
A
M
S
S
C
R
10 L
CPR Motor
S
CF-A
White
Brown
Black
Sump Heater
D
O.D Fan Motor
BK BL
230 V
24 V
BK
OR
Hi
Black
Red
Lo
F-1
CR
MS-1
BR RD
DFC
Black
BR RD
RD
BR RD
1
2
3
4
5
6
7

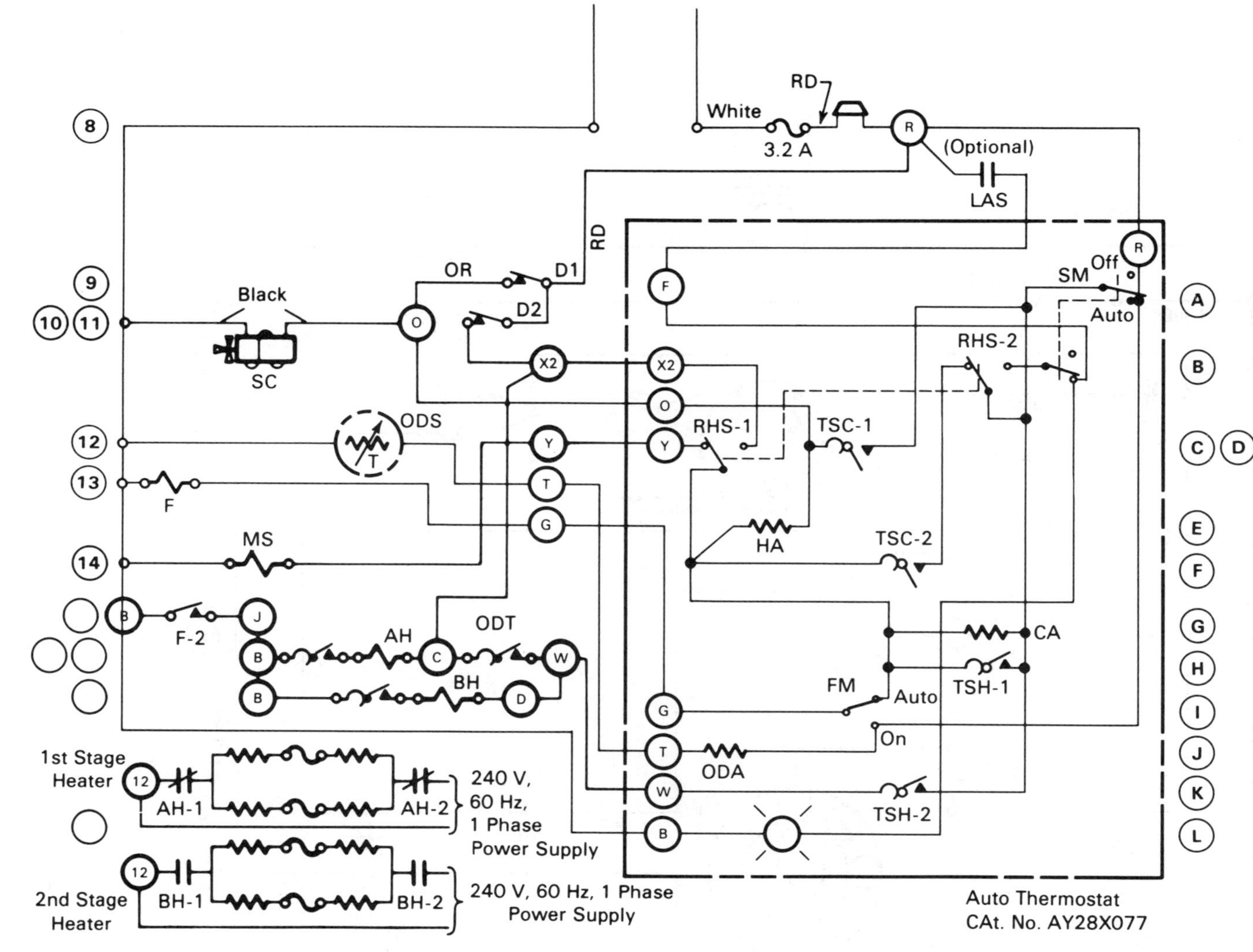

Figure 5–2 Typical Weathertron heat pump schematic diagram, heating and defrost cycles. (Courtesy of The Trane Co., Dealer Products Group).

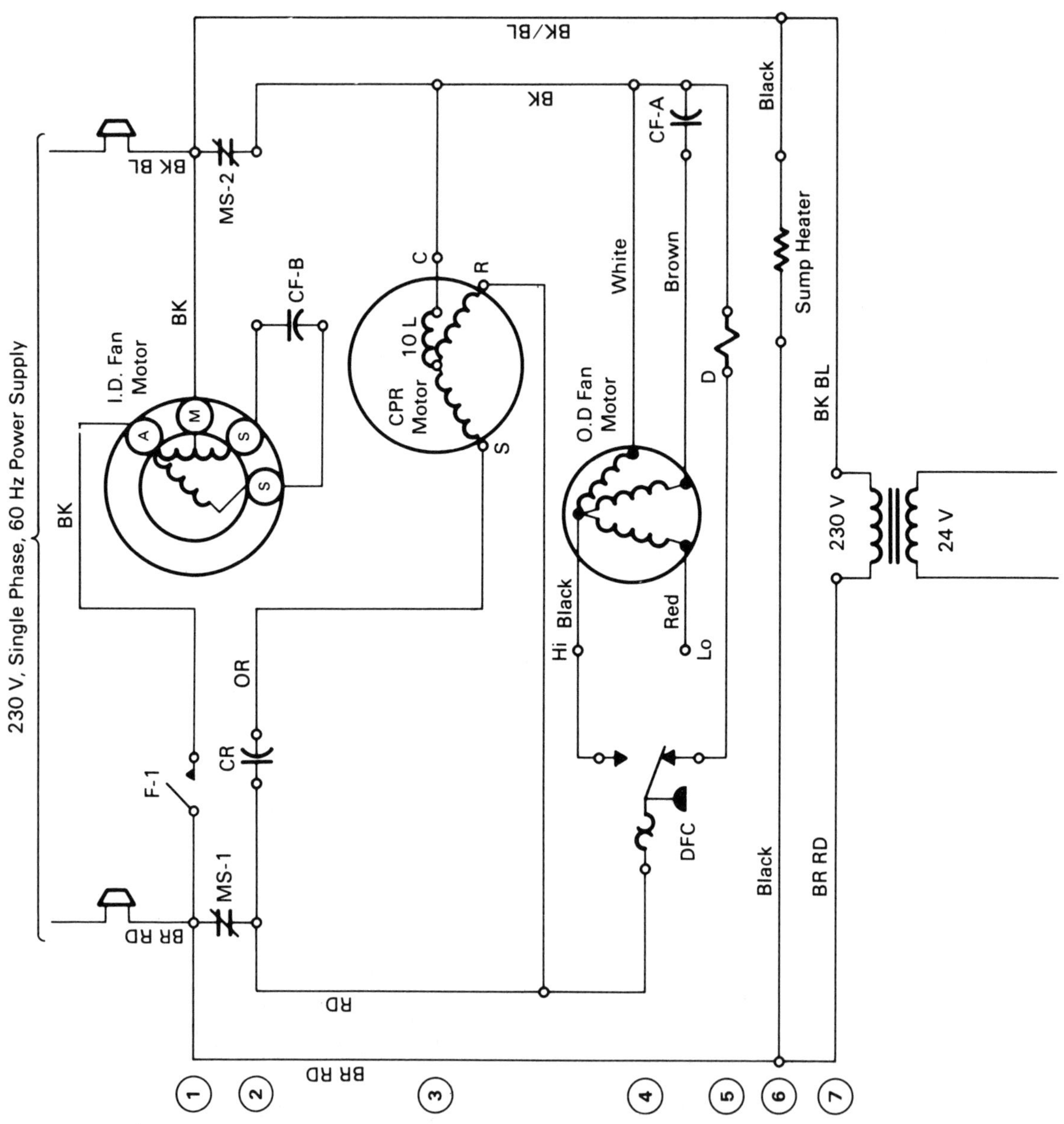
230 V, Single Phase, 60 Hz Power Supply
BK BL
BR RD
MS-2
MS-1
BK
I.D. Fan Motor
A
M
S
S
CF-B
BK
F-1
CR
OR
CPR Motor
10 L
C
R
S
BK/BL
BK
RD
BR RD
CF-A
White
Brown
O.D Fan Motor
D
Hi Black
Red
Lo
DFC
Black
Sump Heater
Black
BK BL
BR RD
230 V
24 V
1
2
3
4
5
6
7

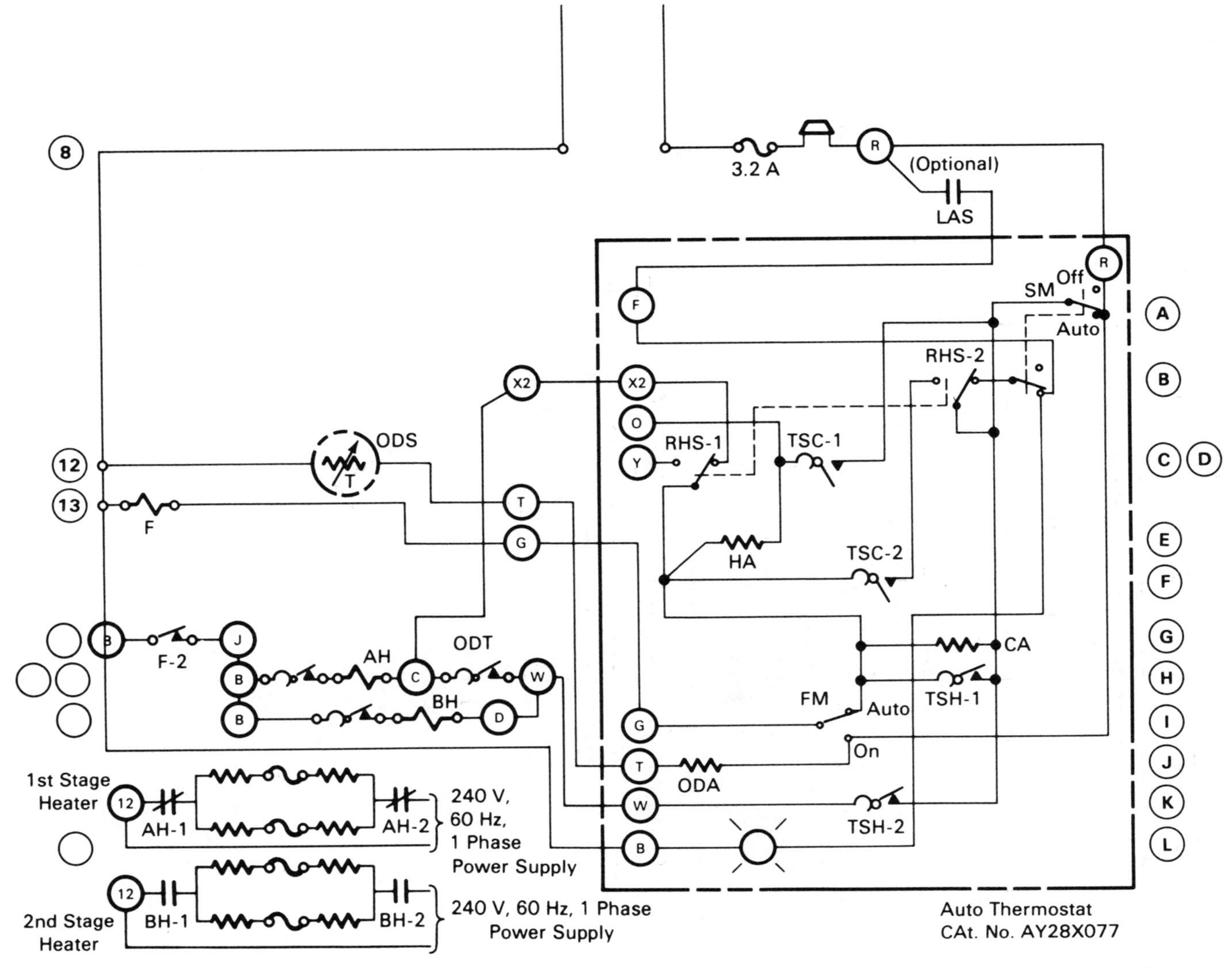

Figure 5–3 Typical Weathertron heat pump schematic diagram, heating and defrost cycles. (Courtesy of The Trane Co., Dealer Products Group.)

5.5 WEATHERTRON CONDENSED VERSION ELECTRICAL SEQUENCE

Thermostat set for cooling: disconnect turned on. See Figure 5-4.

(G) Cooling anticipator energized.
(C) TSC-1 makes 1.5° F (0.84° C) before thermostat calls.
(10) S.O.V. coil energized.
(F) Thermostat calls for cooling TSC-2 makes.
(C) The Y circuit makes.
(I) The G circuit makes.
(14) MS coil energizes.
(13) Fan coil energizes.
(2) MS-1 and MS-2 contacts make.
(1) F-1 contacts make.
(15) F-2 contacts make.

Thermostat set for heating season. See Figure 5-5.

First-stage heating:

(H) TSH-1 contacts make.
(14) MS coil energizes.
(13) F-coil energizes.
(E) Heat anticipator circuit energizes.
(2) MS-1 and MS-2 contacts make.
(1) F-1 fan contacts make.
(15) F-2 fan contacts make.

Second-Stage Heating:

(K) TSH-2 contacts make.
(17) ODT closed.
(16) Supplementary heaters coil A.H. energizes.
(18) Supplementary heaters coil B.H. energizes.

Defrost:

(4) DFC switches to defrost position.
(4) OD fan off.
(5) D relay energizes.
(9) D-1 contacts make.
(11) D-2 contacts make.
(10) S.O.V. energized.
(16) AH coil energized.

Defrost Termination:

(4) DFC switches to normal position.
(4) OD fan starts.
(5) D relay de-energizes.
(9) D-1 opens, de-energizing S.O.V.
(11) D-2 opens, de-energizing AH.

Electric Heat: Manually switched to electric heat position:

(C) RSH-1 disconnects Y and makes X^2 circuit.
(B) RSH-2 disconnects TSC-2 and makes light circuit L.

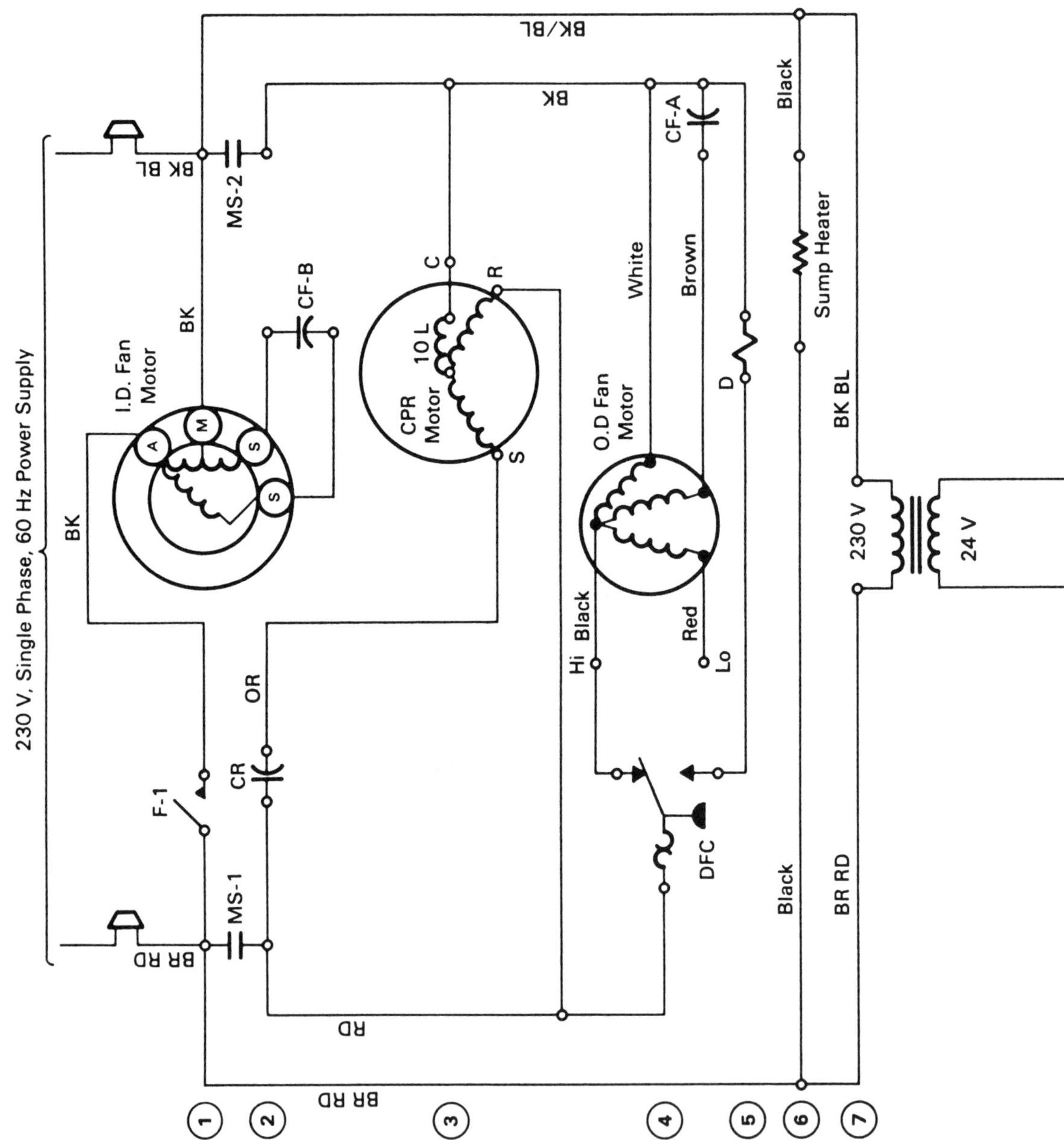
230 V, Single Phase, 60 Hz Power Supply
BK/BL
BK
BK BL
MS-2
CF-A
Black
Sump Heater
I.D. Fan Motor
A
M
S
S
CF-B
C
R
10 L
CPR Motor
S
White
Brown
D
O.D Fan Motor
BK BL
230 V
24 V
Hi Black
Red
Lo
BK
F-1
CR
OR
DFC
MS-1
BR RD
RD
Black
BR RD
BR RD
1
2
3
4
5
6
7

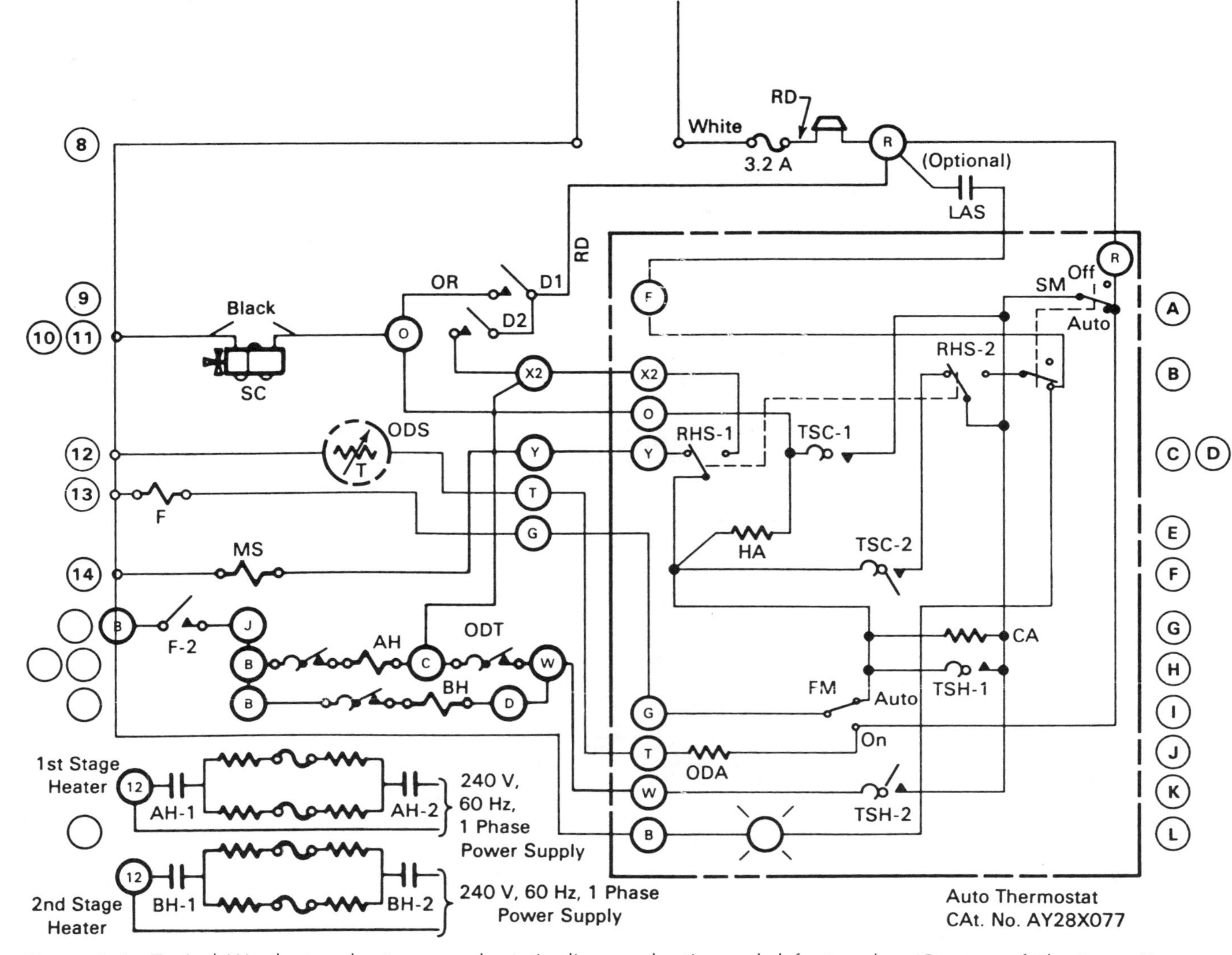

Figure 5–5 Typical Weathertron heat pump schematic diagram, heating and defrost cycles. (Courtesy of The Trane Co., Dealer Products Group.)

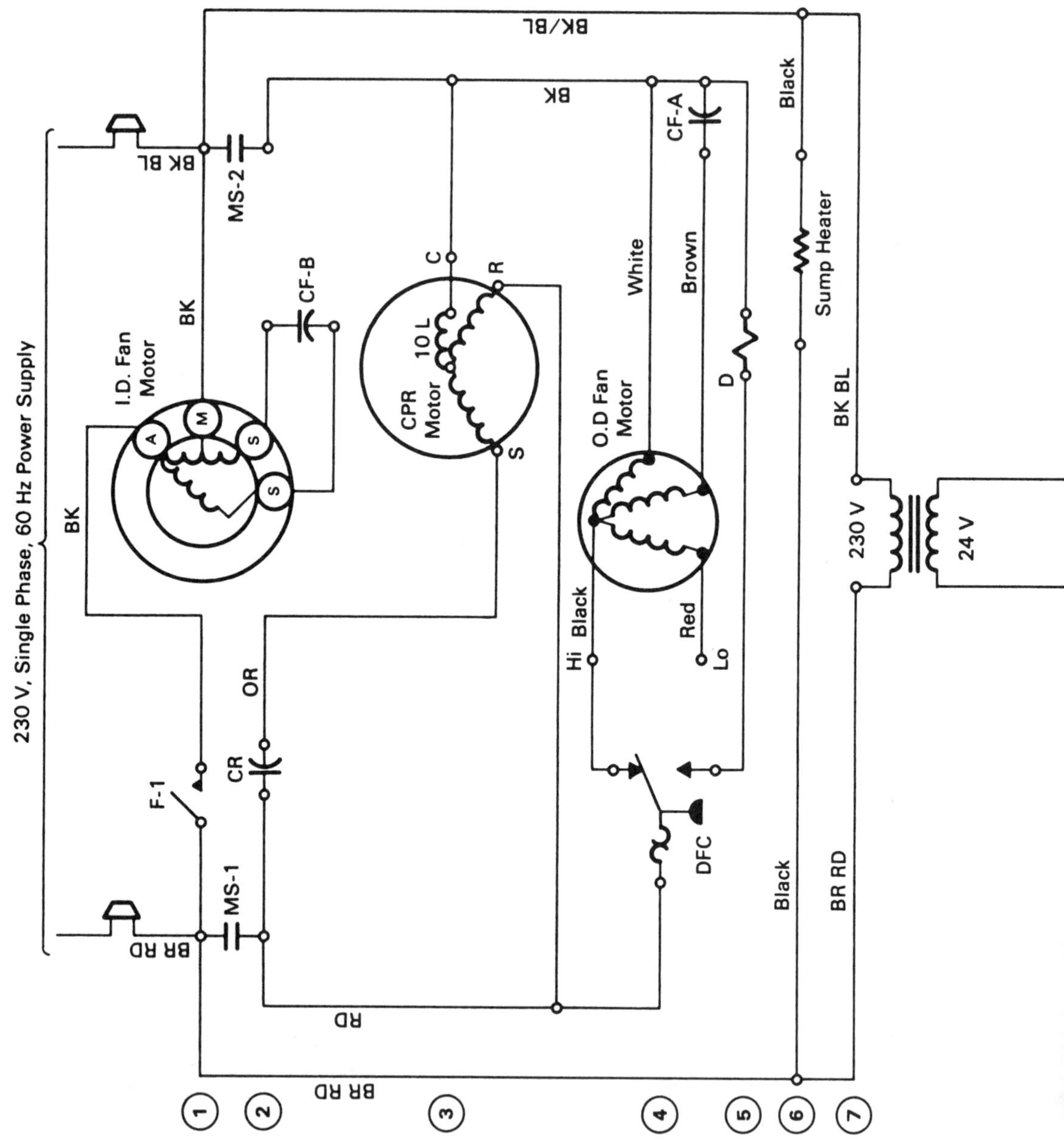
230 V, Single Phase, 60 Hz Power Supply
BK BL
BR RD
BK/BL
BK
MS-2
MS-1
I.D. Fan Motor
A
M
S
S
CF-B
C
R
S
10 L
CPR Motor
White
Brown
CF-A
Black
Sump Heater
D
BK BL
O.D Fan Motor
Hi Black
Red Lo
OR
CR
F-1
DFC
Black
BR RD
RD
230 V
24 V
1
2
3
4
5
6
7

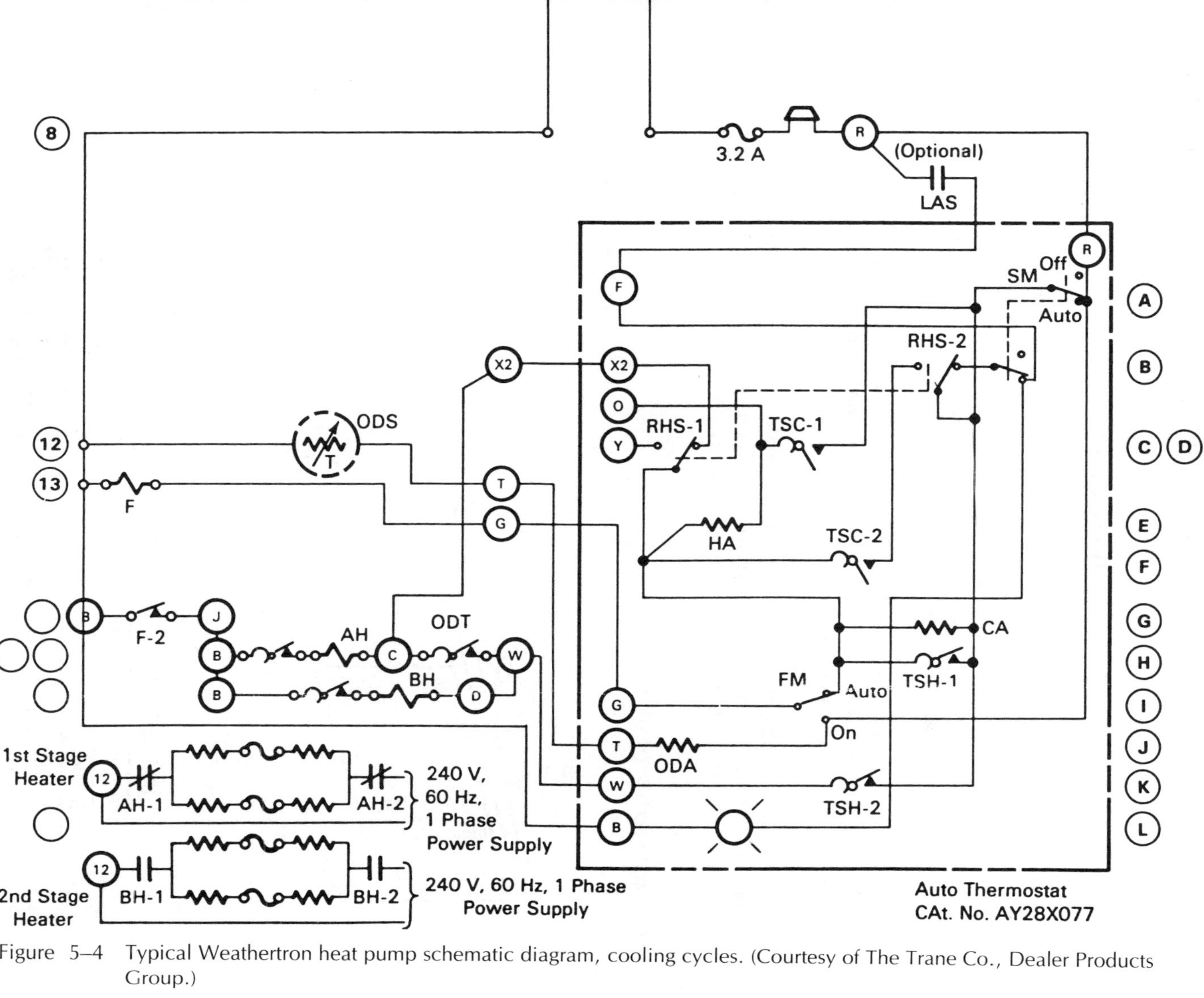

Figure 5–4 Typical Weathertron heat pump schematic diagram, cooling cycles. (Courtesy of The Trane Co., Dealer Products Group.)

(2) INSTALLATION INSTRUCTIONS

IMPORTANT - These instructions and any instructions packaged with any separate equipment required to make up the entire system should be carefully read before beginning the installation. Note particularly starting procedure and any tags and/or labels attached to the equipment.

GENERAL - These instructions explain how to install the air cooled condensing unit (outdoor unit), the matching indoor evaporator and the inter-connecting refrigerant tubing.

The installation should be made in accordance to national codes and/or local code and, authorities having jurisdiction should be consulted before the installation is made.

dimensions (in inches)

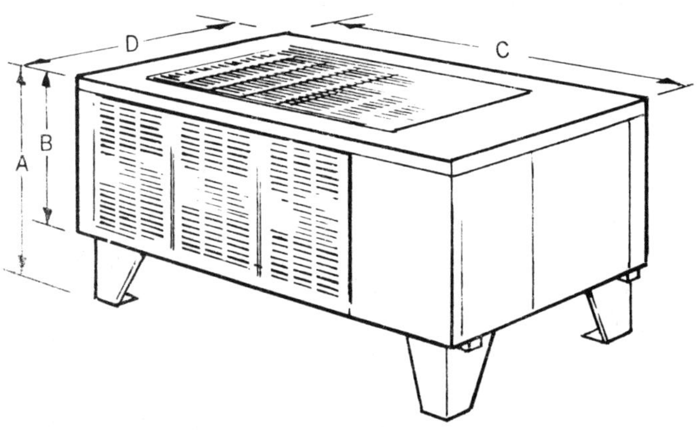

UNIT MODEL NO.	A	B	C	D
CTD-903	30⅜	22⅜	69¼	37¾
CTD-904	30⅜	22⅜	69¼	37¾
CTD-1203	30⅜	22⅜	69¼	37¾
CTD-1204	30⅜	22⅜	69¼	37¾

* Courtesy Heatwave® International, Inc.

These wiring diagrams will be found helpful when servicing the electrical components. These are the same wiring diagrams that are provided with the unit.

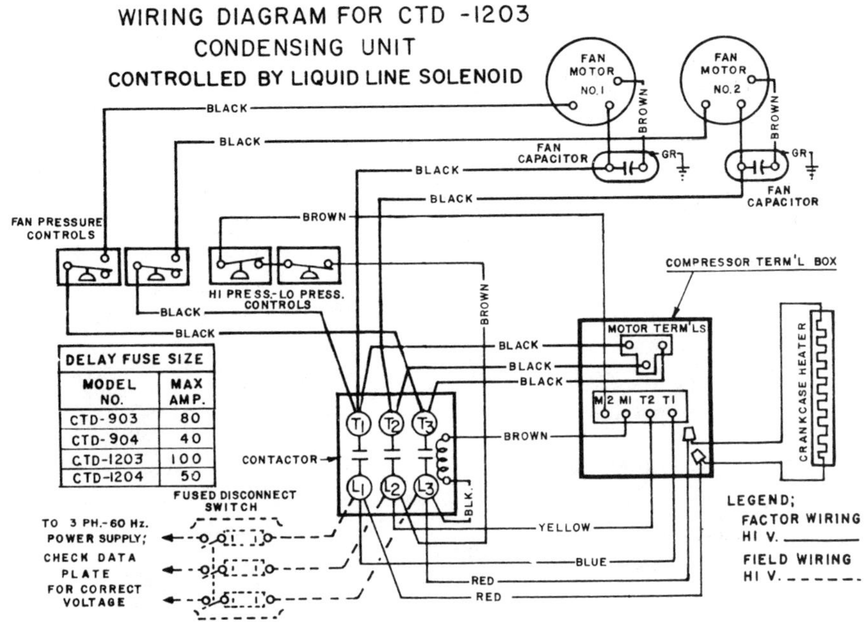

DELAY FUSE SIZE

MODEL NO.	MAX AMP.
CTD-903	80
CTD-904	40
CTD-1203	100
CTD-1204	50

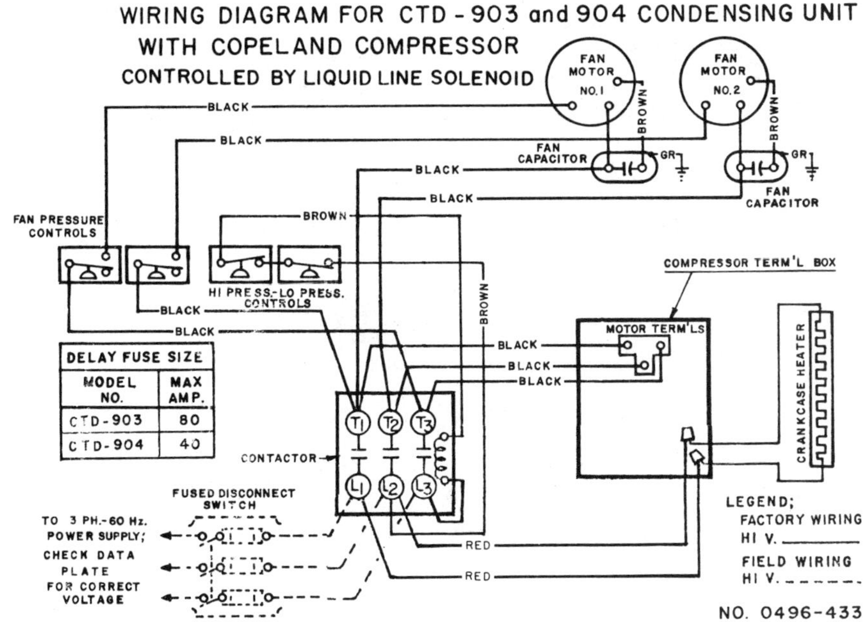

DELAY FUSE SIZE

MODEL NO.	MAX AMP.
CTD-903	80
CTD-904	40

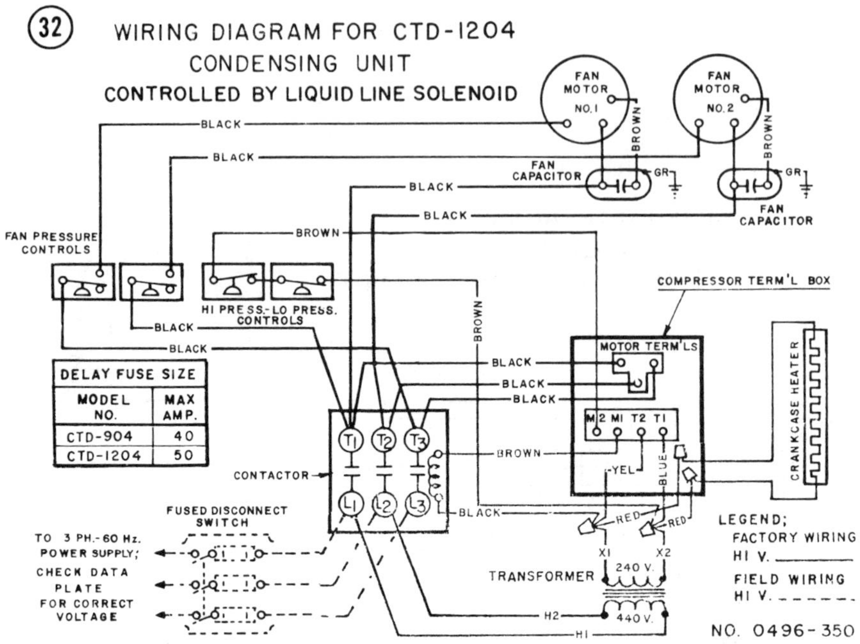

DELAY FUSE SIZE	
MODEL NO.	MAX AMP.
CTD-904	40
CTD-1204	50

WIRING ARRANGENT FOR "CTD" Units W/ Fan Pressure Control, Solenoid & TC Panel at Blower-Coil Unit

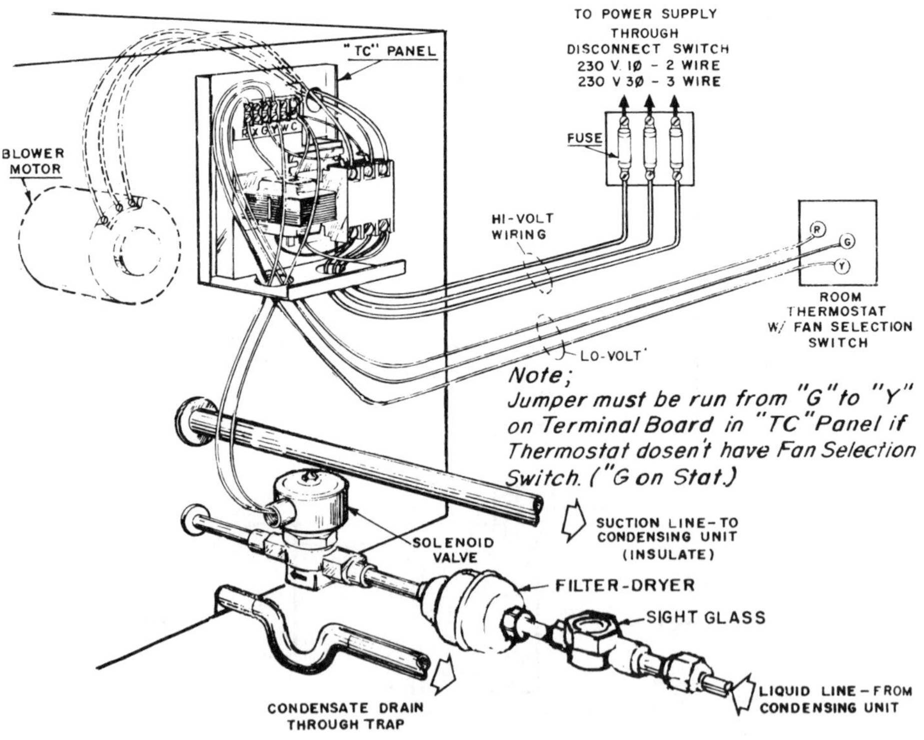

PART No. 0496-369

(34)

WIRING DIAGRAM FOR TC-2 PANEL WITH 1Ø AIR HANDLER AND CTD-903 and 1203 COND. UNIT

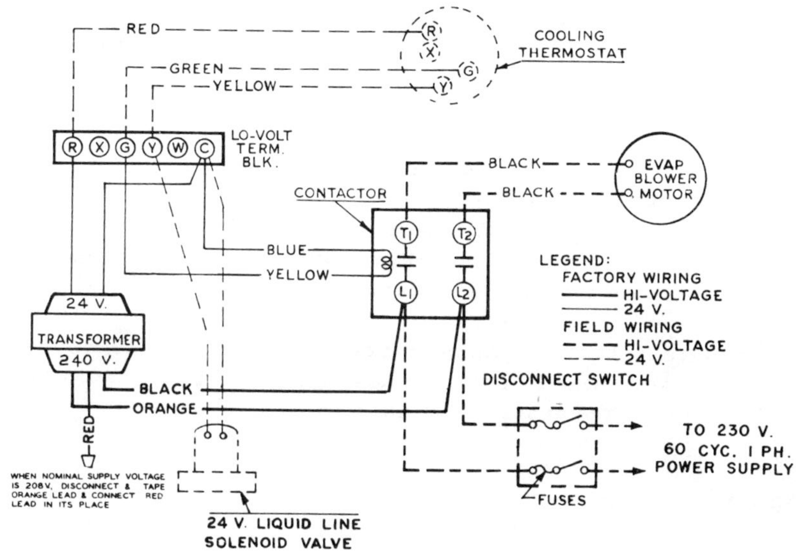

WIRING DIAGRAM FOR TC-3 PANEL (240 V.) WITH 3Ø AIR HANDLER AND CTD-903 and 1203 COND. UNIT

OR TC-4 PANEL (440 V.) WITH 3Ø AIR HANDLER AND CTD-904 and 1204 COND. UNIT

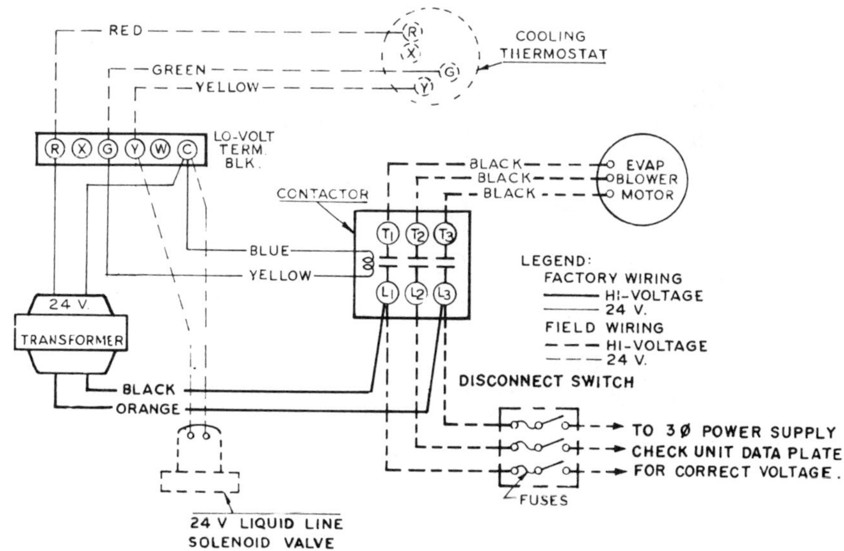

INSTALLATION INSTRUCTIONS

IMPORTANT - These instructions and any instructions packaged with any separate equipment required to make up the entire system should be carefully read before beginning the installation. Note particularly "Connecting Quick-Connect Couplings, Starting Procedure and any tags and/or labels attached to the equipment.

GENERAL - These instructions explain how to install the pre-charged air cooled condensing unit (outdoor unit), the matching indoor pre-charged evaporator (furnace coil) and the interconnecting pre-charged refrigerant tubing. The same general instructions apply to the field-charged units as well.

The installation should be made in accordance to national codes and/or local code and authorities having jurisdiction should be consulted before the installation is made.

dimensions (in inches)

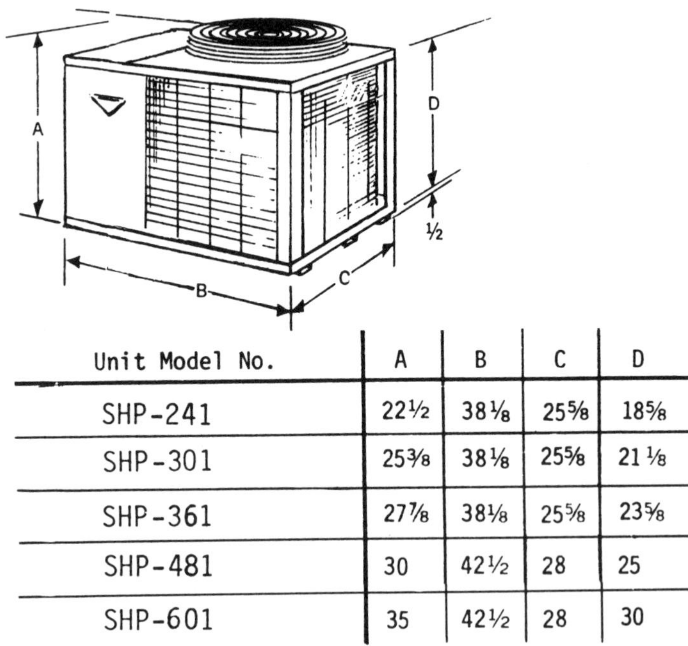

Unit Model No.	A	B	C	D
SHP-241	22½	38⅛	25⅝	18⅝
SHP-301	25⅜	38⅛	25⅝	21⅛
SHP-361	27⅞	38⅛	25⅝	23⅝
SHP-481	30	42½	28	25
SHP-601	35	42½	28	30

Figure 1

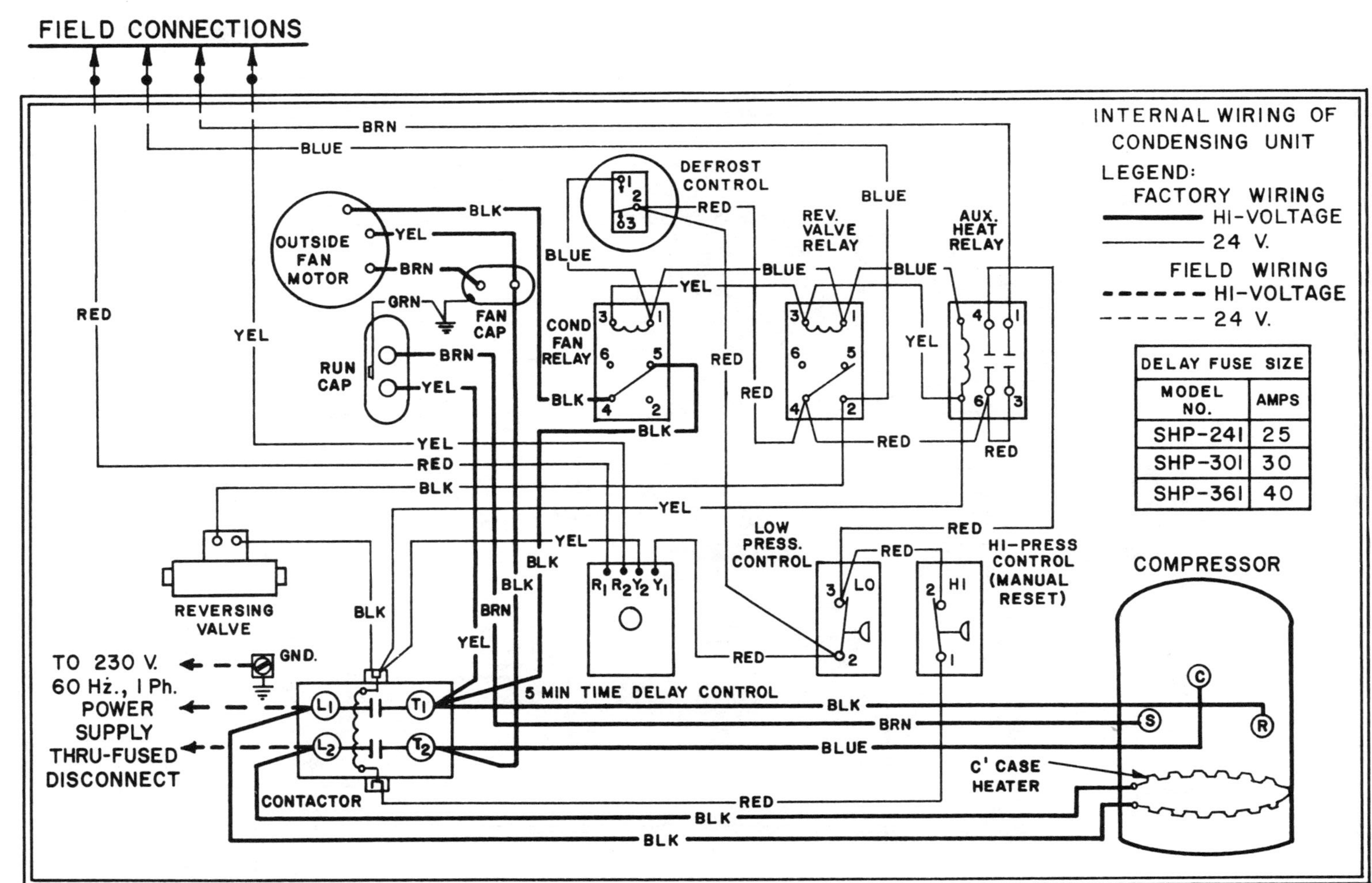

DELAY FUSE SIZE	
MODEL NO.	AMPS
SHP-241	25
SHP-301	30
SHP-361	40

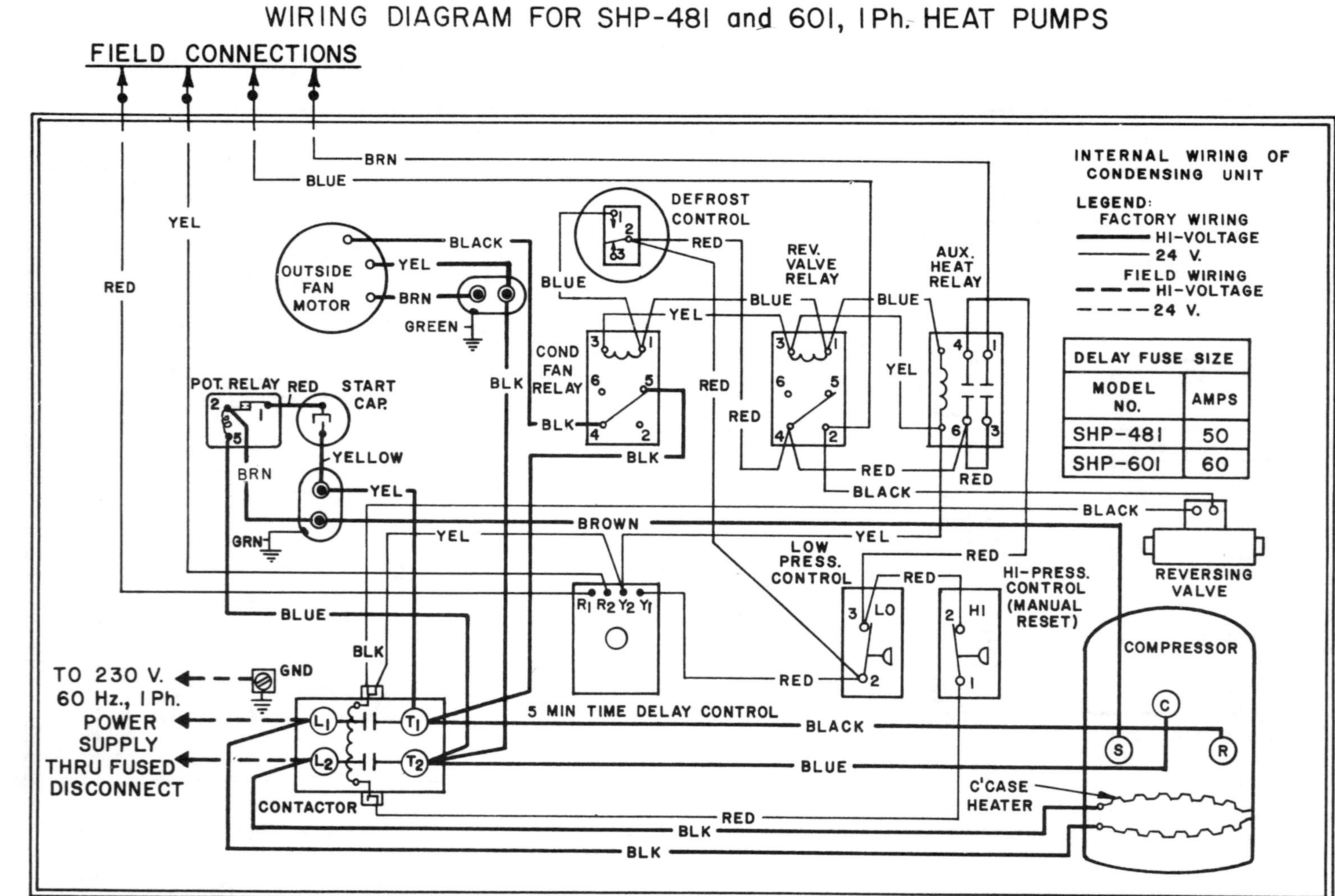
WIRING DIAGRAM FOR SHP-481 and 601, 1Ph. HEAT PUMPS
FIELD CONNECTIONS
INTERNAL WIRING OF CONDENSING UNIT
LEGEND:
FACTORY WIRING
HI-VOLTAGE
24 V.
FIELD WIRING
HI-VOLTAGE
24 V.
DELAY FUSE SIZE
MODEL NO.
AMPS
SHP-481
50
SHP-601
60
BRN
BLUE
YEL
RED
OUTSIDE FAN MOTOR
BLACK
GREEN
DEFROST CONTROL
REV. VALVE RELAY
AUX. HEAT RELAY
COND FAN RELAY
BLK
POT. RELAY
START CAP.
YELLOW
BROWN
LOW PRESS. CONTROL
HI-PRESS. CONTROL (MANUAL RESET)
LO
HI
REVERSING VALVE
COMPRESSOR
5 MIN TIME DELAY CONTROL
R1 R2 Y2 Y1
GRN
GND
TO 230 V. 60 Hz., 1Ph. POWER SUPPLY THRU FUSED DISCONNECT
CONTACTOR
L1
L2
T1
T2
C'CASE HEATER
S
C
R
Nº 0496-504

HOOK-UP WIRING DIAGRAM FOR SHP- 301 & 361
with HPM- 15 with AHM- 18-42

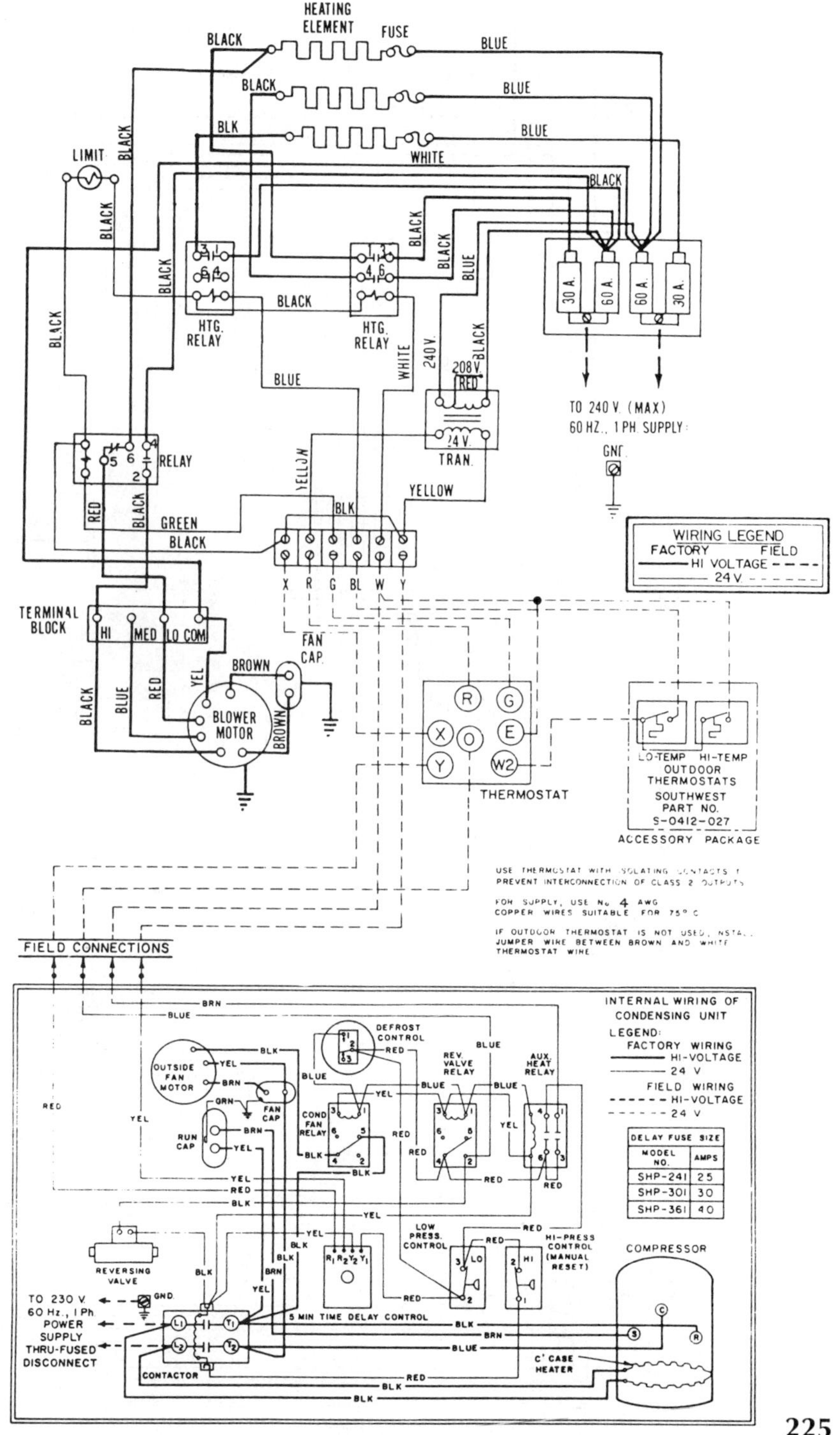

0496-524

HOOK-UP WIRING DIAGRAM FOR SHP- 241
with HPM- 15 with AHM- 15-24

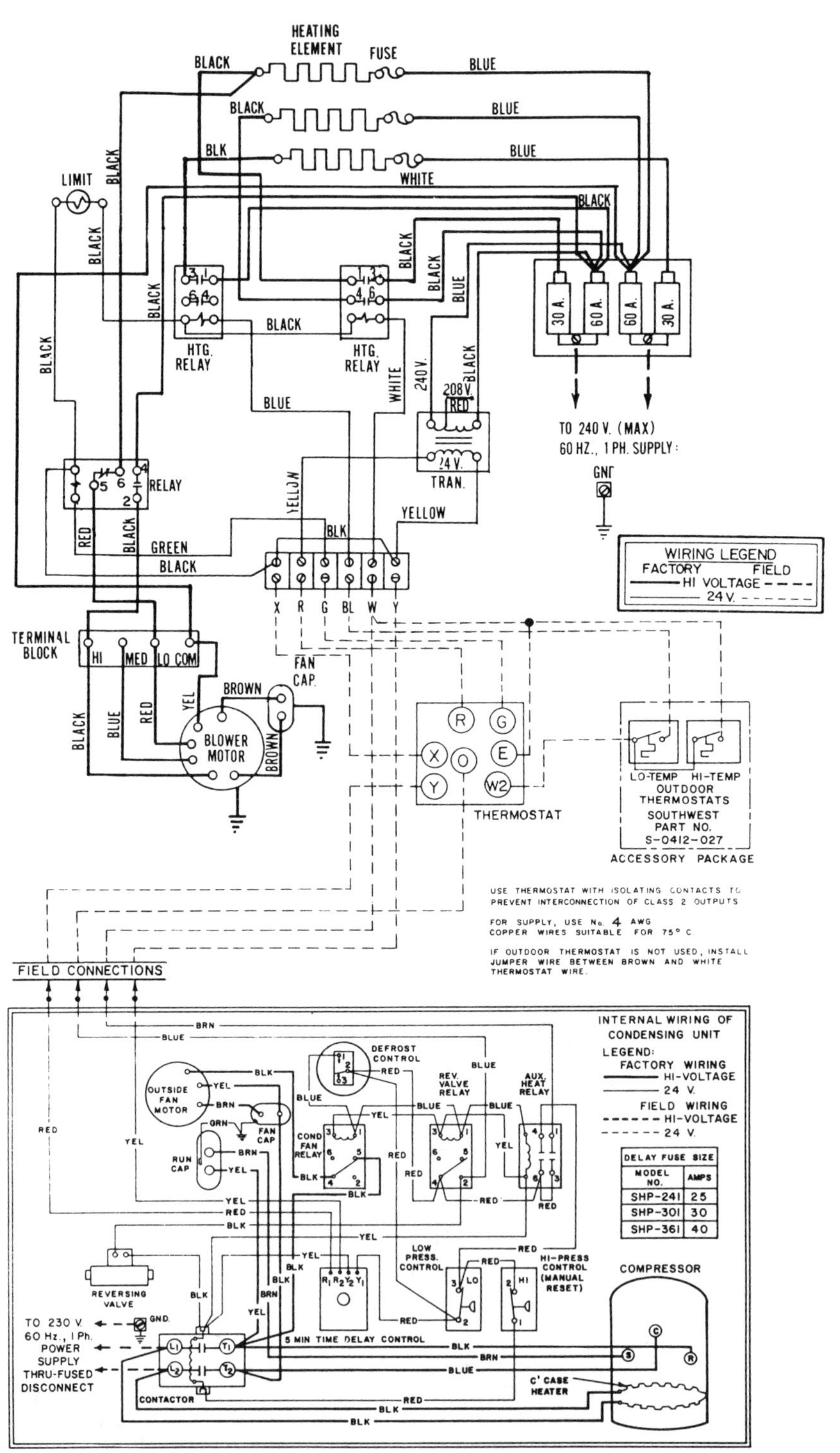

HOOK-UP WIRING DIAGRAM FOR SHP-301 & 361
with HPM-10 with AHM-18-42

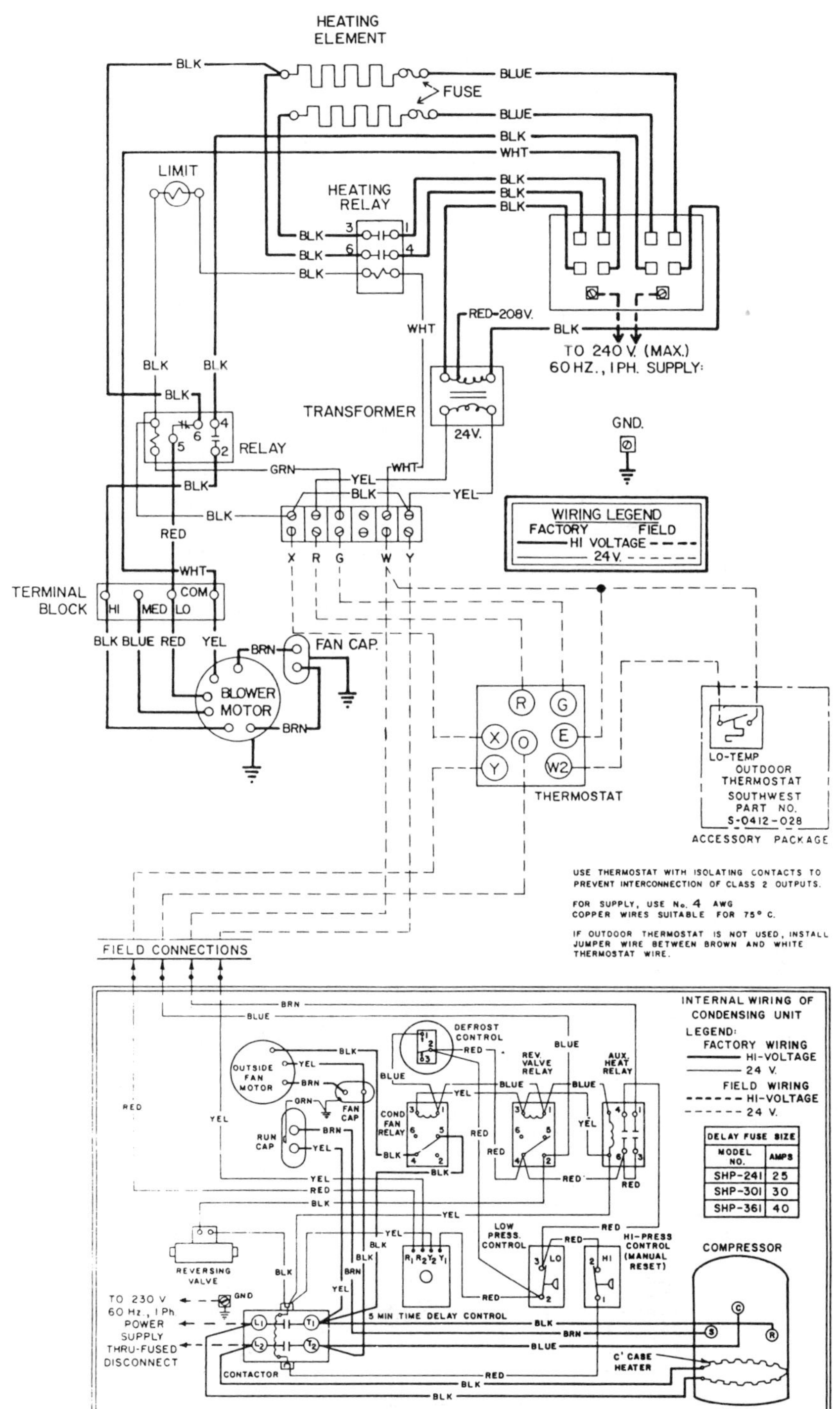

0496-522

HOOK-UP WIRING DIAGRAM FOR SHP-241
with HPM-7.5&10 with AHM-15-24

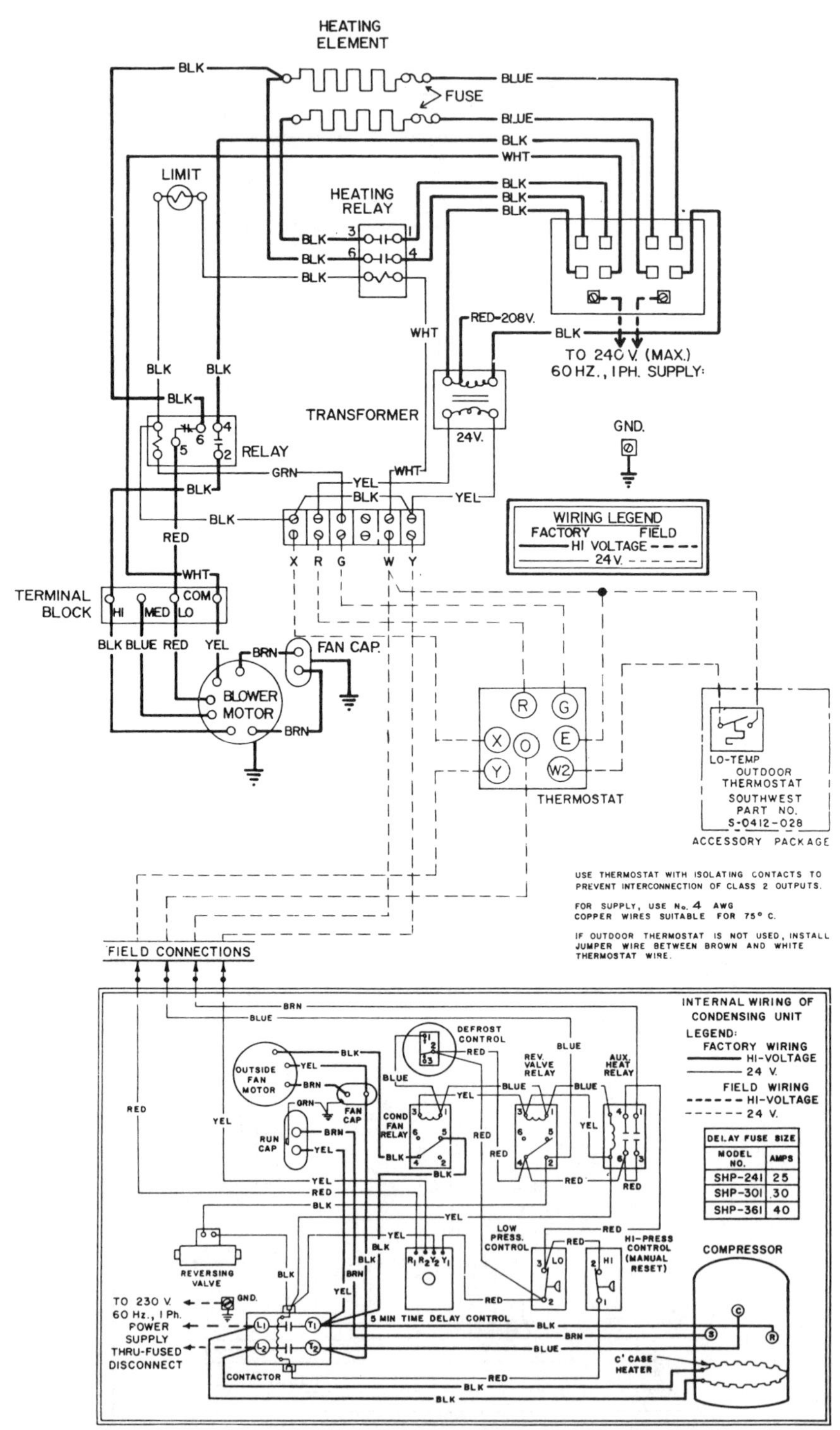

0496-521

HOOK-UP WIRING DIAGRAM FOR SHP-301 & 361
with HPM- 5 with AHM- 18-42

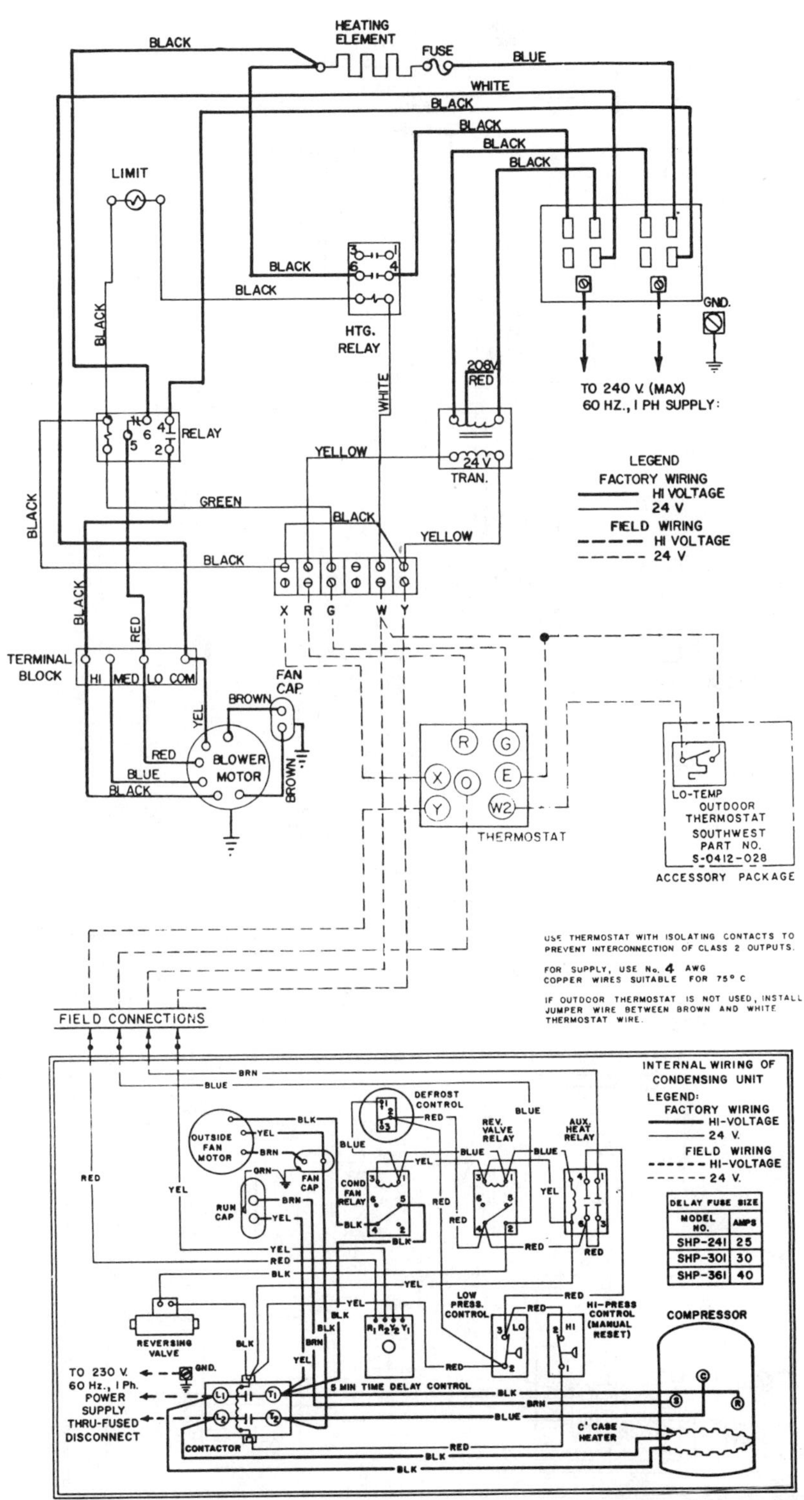

DELAY FUSE SIZE	
MODEL NO.	AMPS
SHP-241	25
SHP-301	30
SHP-361	40

0496-520

HOOK-UP WIRING DIAGRAM FOR SHP- 241
with HPM- 3 & 5 with AHM- 15-24

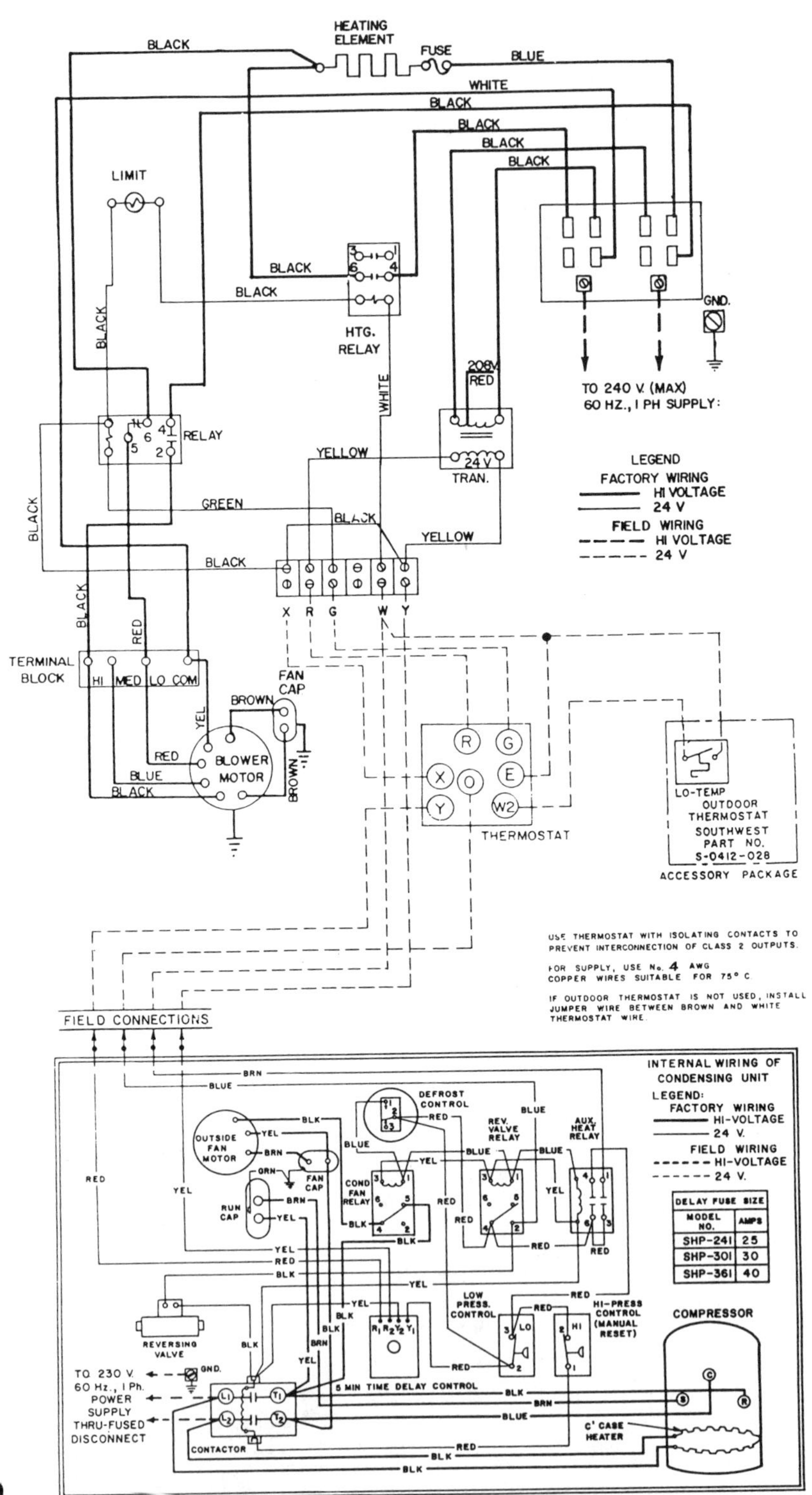

DELAY FUSE SIZE	
MODEL NO.	AMPS
SHP-241	25
SHP-301	30
SHP-361	40

0496-519

$1\frac{1}{2}$ to $3\frac{1}{2}$ Ton
Single Phase

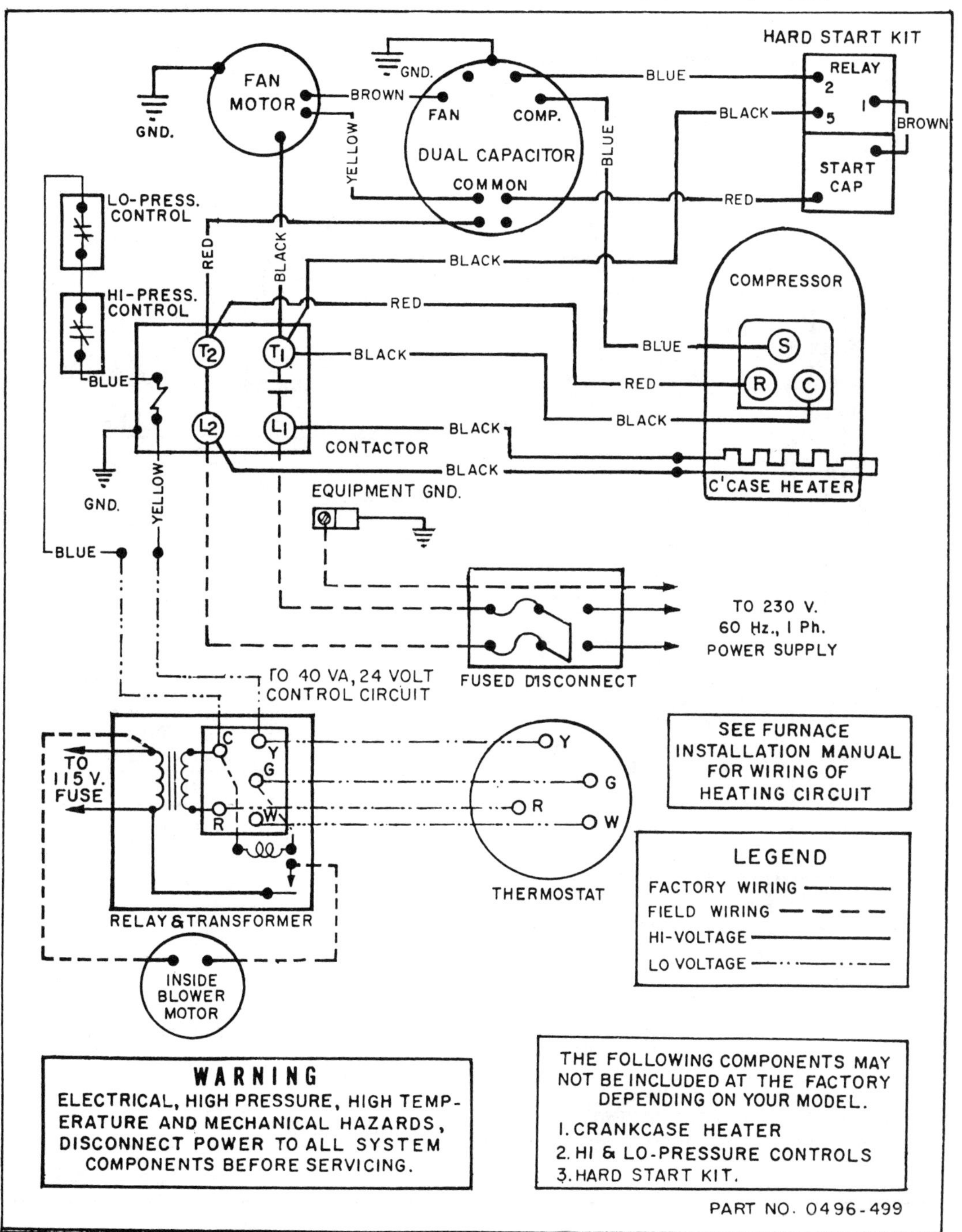

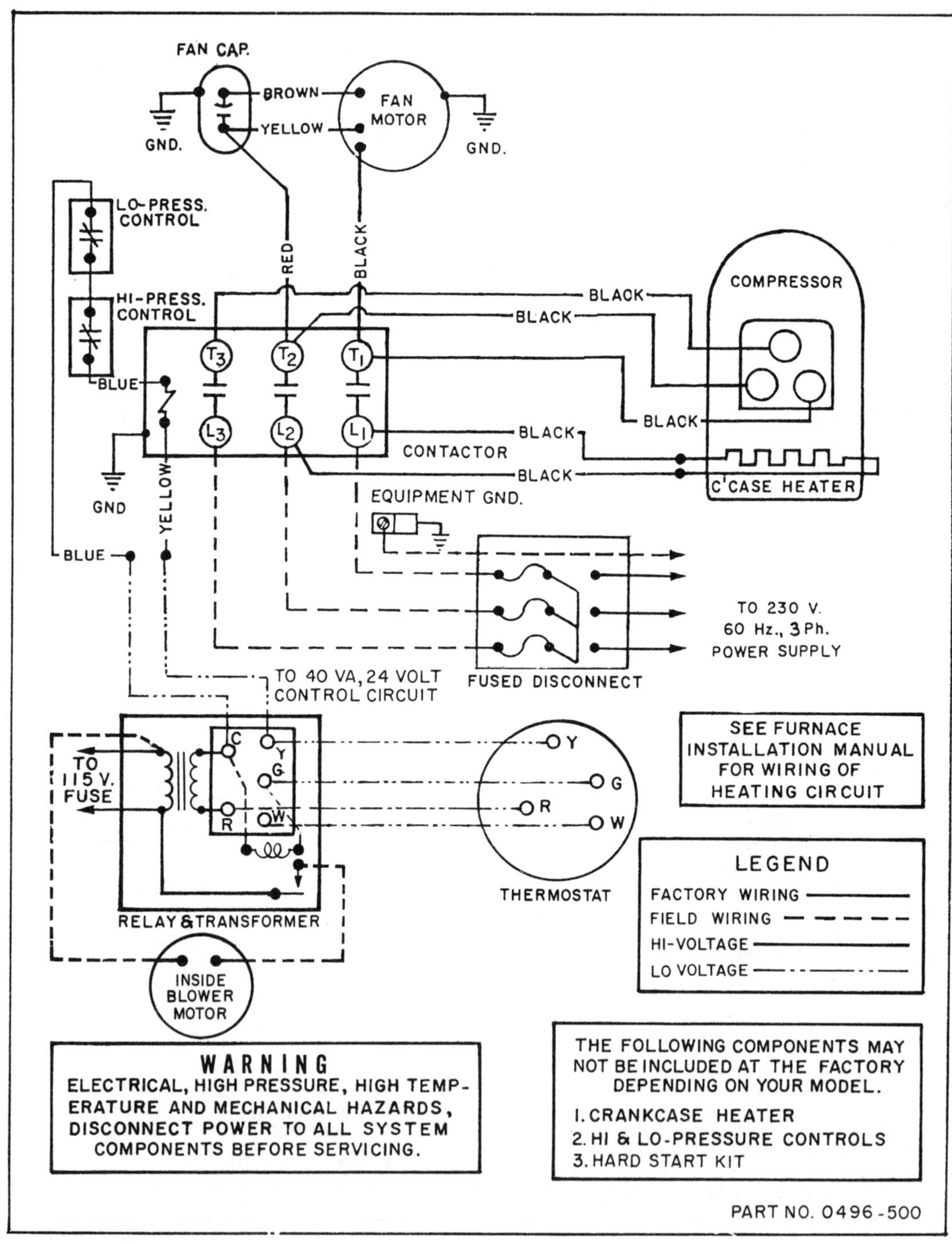
3 Phase
COMPLETE WIRING WITH ALL ACCESSORIES
FAN CAP.
BROWN
YELLOW
FAN MOTOR
GND.
GND.
LO-PRESS. CONTROL
HI-PRESS. CONTROL
RED
BLACK
BLACK
BLACK
COMPRESSOR
BLUE
T3
T2
T1
L3
L2
L1
BLACK
BLACK
CONTACTOR
BLACK
C'CASE HEATER
GND
YELLOW
EQUIPMENT GND.
BLUE
TO 230 V. 60 Hz., 3 Ph. POWER SUPPLY
TO 40 VA, 24 VOLT CONTROL CIRCUIT
FUSED DISCONNECT
TO 115 V. FUSE
C
Y
G
R
W
Y
G
R
W
SEE FURNACE INSTALLATION MANUAL FOR WIRING OF HEATING CIRCUIT
LEGEND
FACTORY WIRING
FIELD WIRING
HI-VOLTAGE
LO VOLTAGE
THERMOSTAT
RELAY & TRANSFORMER
INSIDE BLOWER MOTOR
WARNING
ELECTRICAL, HIGH PRESSURE, HIGH TEMPERATURE AND MECHANICAL HAZARDS, DISCONNECT POWER TO ALL SYSTEM COMPONENTS BEFORE SERVICING.
THE FOLLOWING COMPONENTS MAY NOT BE INCLUDED AT THE FACTORY DEPENDING ON YOUR MODEL.
1. CRANKCASE HEATER
2. HI & LO-PRESSURE CONTROLS
3. HARD START KIT
PART NO. 0496-500

4 & 5 Ton Units
Single Phase

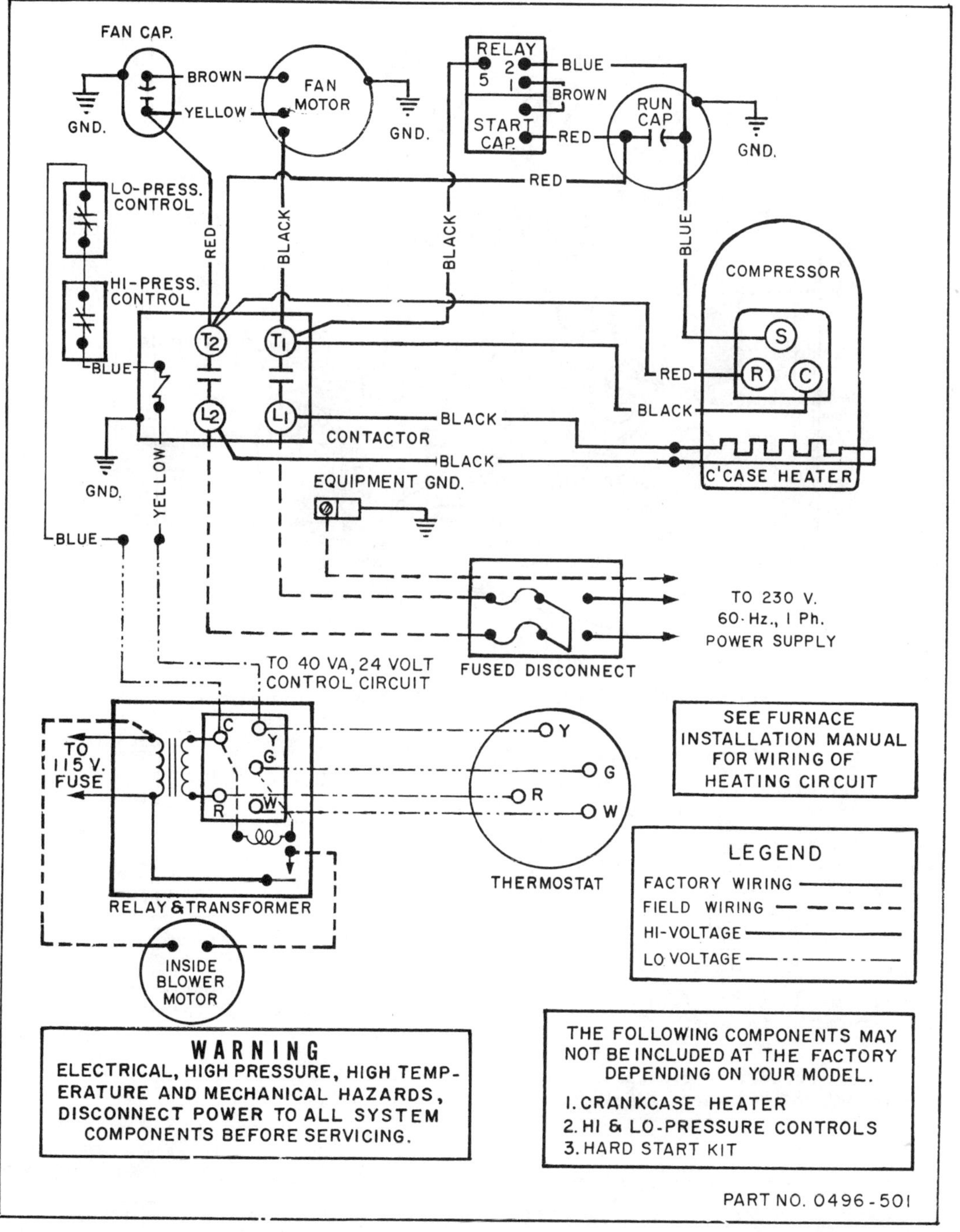

INSTALLATION & WIRING INSTRUCTIONS FOR "SRK" SERIES START RELAY & CAPACITOR KIT

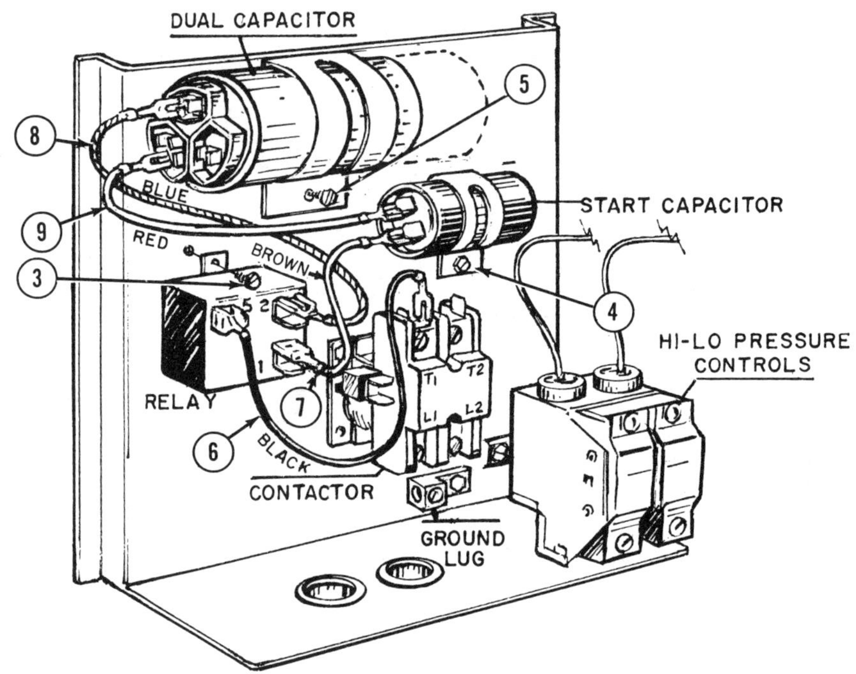

- INSTRUCTIONS -

1. Disconnect electrical power to condening unit.
2. Remove the control panel service door.
3. Install the start relay, as shown, using one (1) #8x1/2" sheet metal screw.
4. Install the start capacitor, as shown, using capacitor strap and two (2) #8x1/2" sheet metal screws.
5. Loosen the dual capacitor mounting screw and slide the capacitor to the left side enough to connect wires.
6. Connect the black wire from the T1 contactor terminal to start relay #5 terminal.
7. Connect the brown wire from start relay #1 terminal to the start capacitor.
8. Connect the blue wire from start relay #2 to the "H" (Hermetic or Compressor) terminal on the dual capacitor.
9. Connect the red wire from the start capacitor to the "C" (common) terminal on the dual capacitor.
10. Double check all wiring against the wiring diagram on the control panel service door.
11. Reinstall the dual capacitor in its original position.
12. Reconnect the electrical power and check for proper operation.
13. Reinstall and secure control panel service door.

Part No. 0449-212

5.7 TYPICAL RUUD WIRING DIAGRAMS*

* Courtesy Ruud Air Conditioning Division Rheem Manufacturing Company.

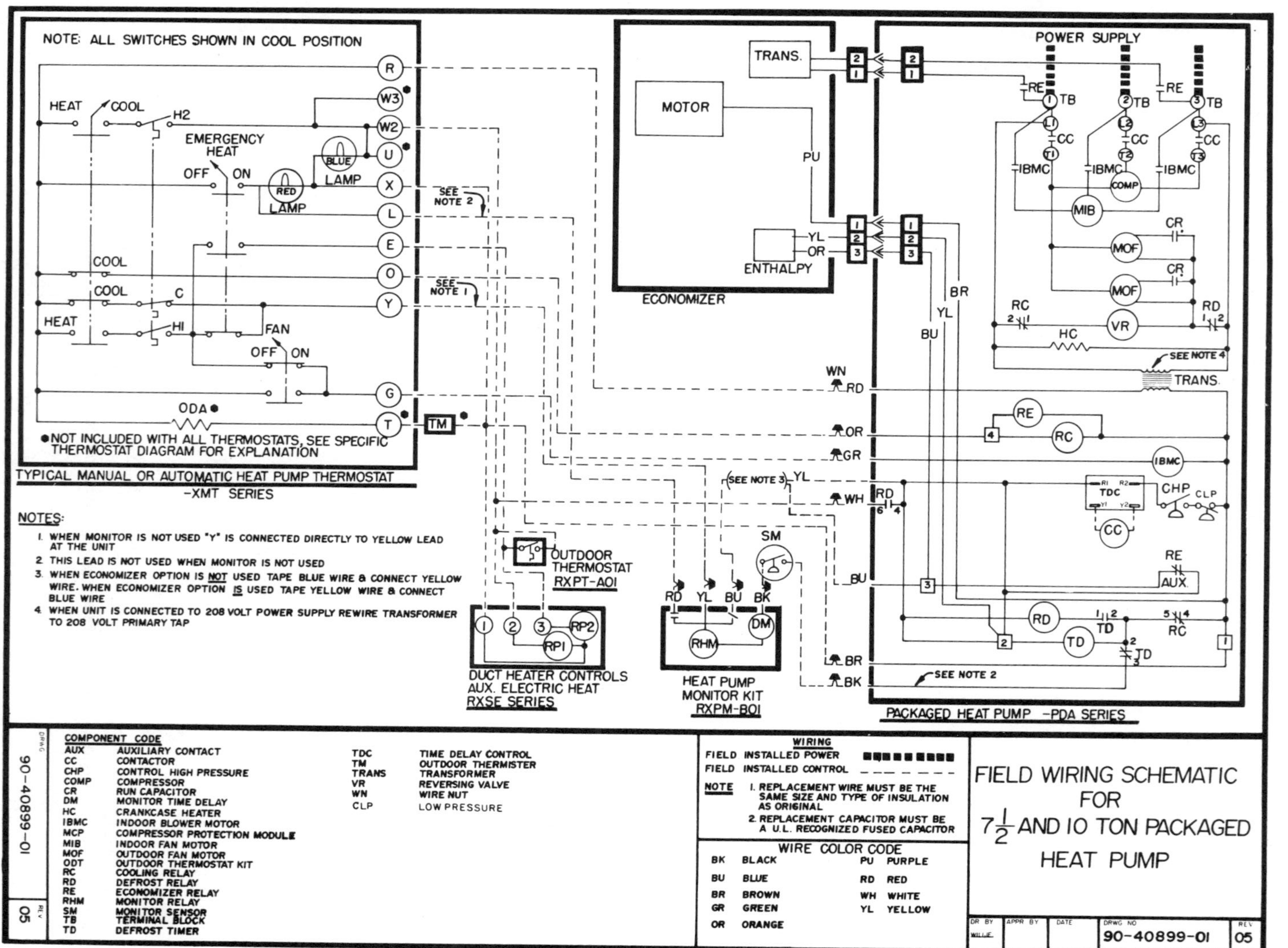
NOTE: ALL SWITCHES SHOWN IN COOL POSITION
HEAT
COOL
H2
EMERGENCY HEAT
OFF
ON
RED LAMP
BLUE LAMP
COOL
COOL
C
HEAT
HI
FAN
OFF
ON
ODA●
R
W3
W2
U
X
L
E
O
Y
G
T
TM
●NOT INCLUDED WITH ALL THERMOSTATS, SEE SPECIFIC THERMOSTAT DIAGRAM FOR EXPLANATION
TYPICAL MANUAL OR AUTOMATIC HEAT PUMP THERMOSTAT -XMT SERIES
SEE NOTE 2
SEE NOTE 1
NOTES:
1. WHEN MONITOR IS NOT USED "Y" IS CONNECTED DIRECTLY TO YELLOW LEAD AT THE UNIT
2. THIS LEAD IS NOT USED WHEN MONITOR IS NOT USED
3. WHEN ECONOMIZER OPTION IS NOT USED TAPE BLUE WIRE & CONNECT YELLOW WIRE. WHEN ECONOMIZER OPTION IS USED TAPE YELLOW WIRE & CONNECT BLUE WIRE
4. WHEN UNIT IS CONNECTED TO 208 VOLT POWER SUPPLY REWIRE TRANSFORMER TO 208 VOLT PRIMARY TAP
TRANS.
MOTOR
PU
YL
OR
ENTHALPY
ECONOMIZER
OUTDOOR THERMOSTAT RXPT-A01
RP2
RP1
DUCT HEATER CONTROLS AUX. ELECTRIC HEAT RXSE SERIES
SM
RD YL BU BK
RHM
DM
HEAT PUMP MONITOR KIT RXPM-B01
WN
RD
OR
GR
(SEE NOTE 3) YL
WH
BU
BR
BK
POWER SUPPLY
RE
TB
CC
IBMC
COMP
MIB
CR
MOF
RC
HC
VR
RD
SEE NOTE 4
TRANS.
RE
RC
IBMC
TDC
CHP
CLP
CC
AUX
TD
SEE NOTE 2
PACKAGED HEAT PUMP -PDA SERIES
DRWG 90-40899-01 REV 05
COMPONENT CODE
AUX AUXILIARY CONTACT
CC CONTACTOR
CHP CONTROL HIGH PRESSURE
COMP COMPRESSOR
CR RUN CAPACITOR
DM MONITOR TIME DELAY
HC CRANKCASE HEATER
IBMC INDOOR BLOWER MOTOR
MCP COMPRESSOR PROTECTION MODULE
MIB INDOOR FAN MOTOR
MOF OUTDOOR FAN MOTOR
ODT OUTDOOR THERMOSTAT KIT
RC COOLING RELAY
RD DEFROST RELAY
RE ECONOMIZER RELAY
RHM MONITOR RELAY
SM MONITOR SENSOR
TB TERMINAL BLOCK
TD DEFROST TIMER
TDC TIME DELAY CONTROL
TM OUTDOOR THERMISTER
TRANS TRANSFORMER
VR REVERSING VALVE
WN WIRE NUT
CLP LOW PRESSURE
WIRING
FIELD INSTALLED POWER
FIELD INSTALLED CONTROL
NOTE 1. REPLACEMENT WIRE MUST BE THE SAME SIZE AND TYPE OF INSULATION AS ORIGINAL
2. REPLACEMENT CAPACITOR MUST BE A U.L. RECOGNIZED FUSED CAPACITOR
WIRE COLOR CODE
BK BLACK
BU BLUE
BR BROWN
GR GREEN
OR ORANGE
PU PURPLE
RD RED
WH WHITE
YL YELLOW
FIELD WIRING SCHEMATIC FOR 7½ AND 10 TON PACKAGED HEAT PUMP
DR BY WILLIE
APPR BY
DATE
DRWG NO 90-40899-01
REV 05

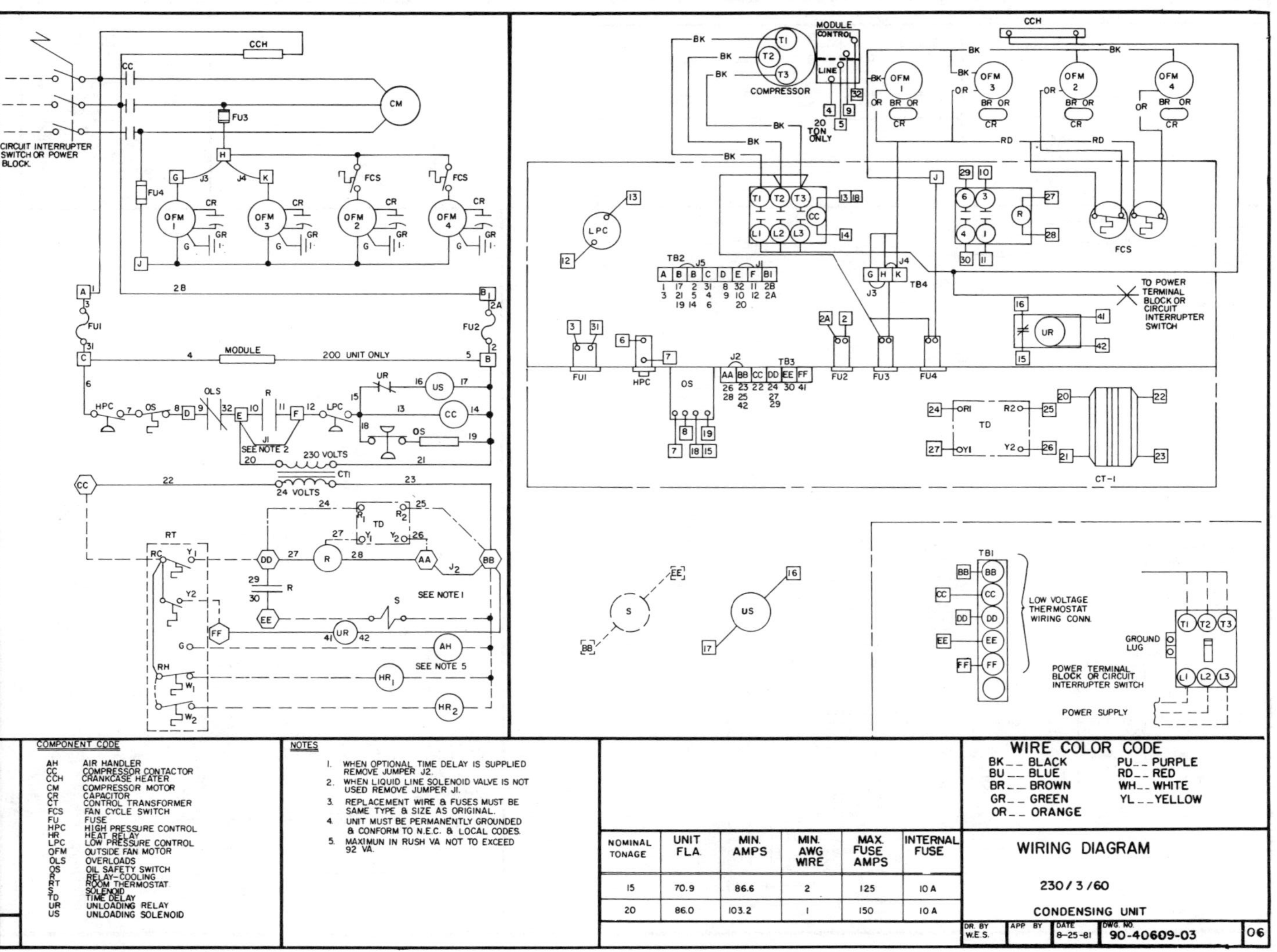

NOMINAL TONAGE	UNIT FLA	MIN. AMPS	MIN. AWG WIRE	MAX. FUSE AMPS	INTERNAL FUSE
15	70.9	86.6	2	125	10 A
20	86.0	103.2	1	150	10 A

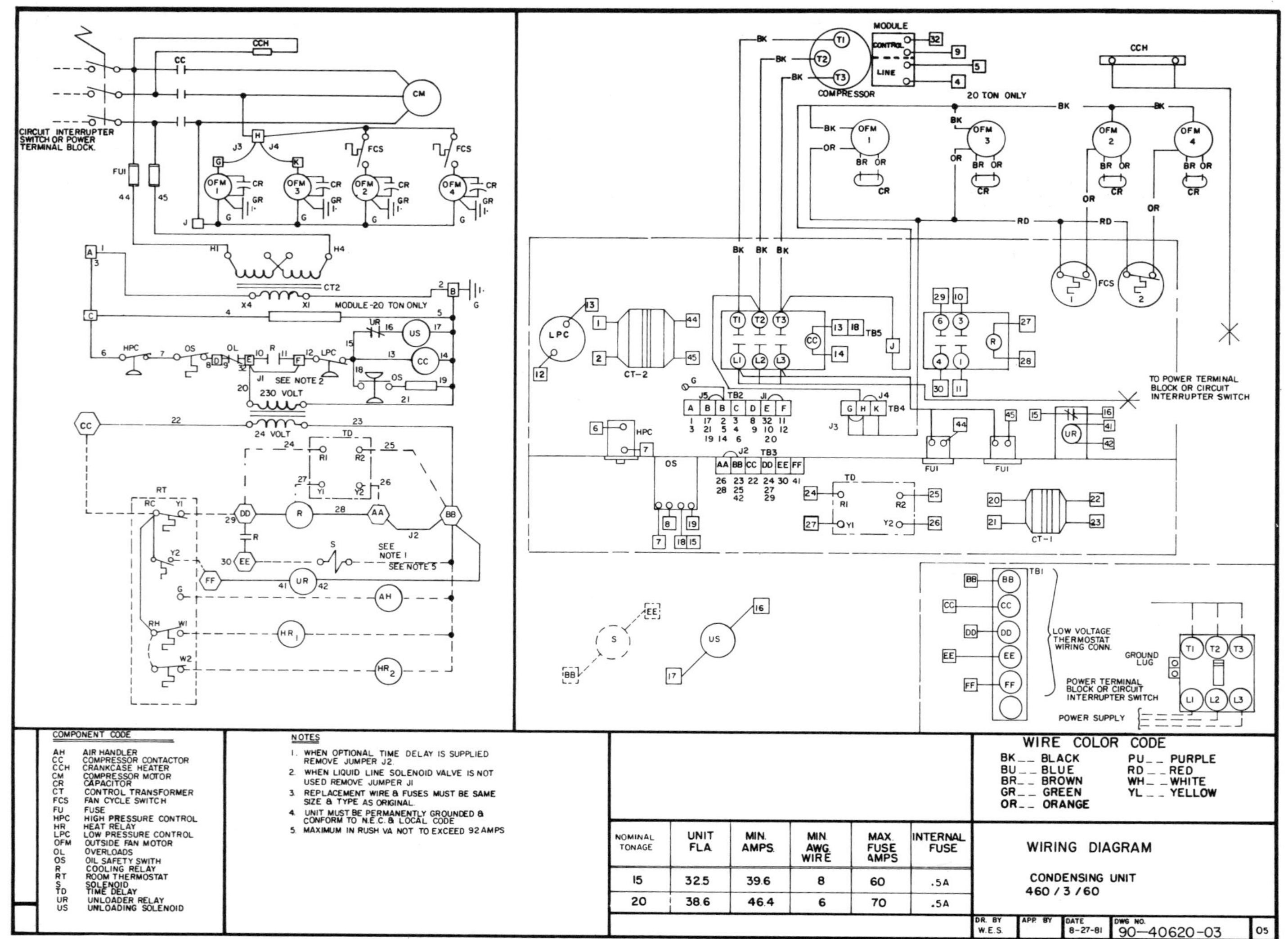
COMPONENT CODE
AH AIR HANDLER
CC COMPRESSOR CONTACTOR
CCH CRANKCASE HEATER
CM COMPRESSOR MOTOR
CR CAPACITOR
CT CONTROL TRANSFORMER
FCS FAN CYCLE SWITCH
FU FUSE
HPC HIGH PRESSURE CONTROL
HR HEAT RELAY
LPC LOW PRESSURE CONTROL
OFM OUTSIDE FAN MOTOR
OL OVERLOADS
OS OIL SAFETY SWITH
R COOLING RELAY
RT ROOM THERMOSTAT
S SOLENOID
TD TIME DELAY
UR UNLOADER RELAY
US UNLOADING SOLENOID
NOTES
1. WHEN OPTIONAL TIME DELAY IS SUPPLIED REMOVE JUMPER J2.
2. WHEN LIQUID LINE SOLENOID VALVE IS NOT USED REMOVE JUMPER J1
3. REPLACEMENT WIRE & FUSES MUST BE SAME SIZE & TYPE AS ORIGINAL.
4. UNIT MUST BE PERMANENTLY GROUNDED & CONFORM TO N.E.C. & LOCAL CODE
5. MAXIMUM IN RUSH VA NOT TO EXCEED 92 AMPS
NOMINAL TONAGE | UNIT FLA | MIN. AMPS. | MIN. AWG. WIRE | MAX. FUSE AMPS | INTERNAL FUSE
15 | 32.5 | 39.6 | 8 | 60 | .5A
20 | 38.6 | 46.4 | 6 | 70 | .5A
WIRE COLOR CODE
BK — BLACK
BU — BLUE
BR — BROWN
GR — GREEN
OR — ORANGE
PU — PURPLE
RD — RED
WH — WHITE
YL — YELLOW
WIRING DIAGRAM
CONDENSING UNIT
460 / 3 / 60
DR. BY W.E.S.
APP. BY
DATE 8-27-81
DWG NO. 90—40620-03
05
CIRCUIT INTERRUPTER SWITCH OR POWER TERMINAL BLOCK.
MODULE-20 TON ONLY
SEE NOTE 2
230 VOLT
24 VOLT
SEE NOTE 1
SEE NOTE 5
COMPRESSOR
MODULE
CONTROL
LINE
20 TON ONLY
TO POWER TERMINAL BLOCK OR CIRCUIT INTERRUPTER SWITCH
LOW VOLTAGE THERMOSTAT WIRING CONN.
GROUND LUG
POWER TERMINAL BLOCK OR CIRCUIT INTERRUPTER SWITCH
POWER SUPPLY

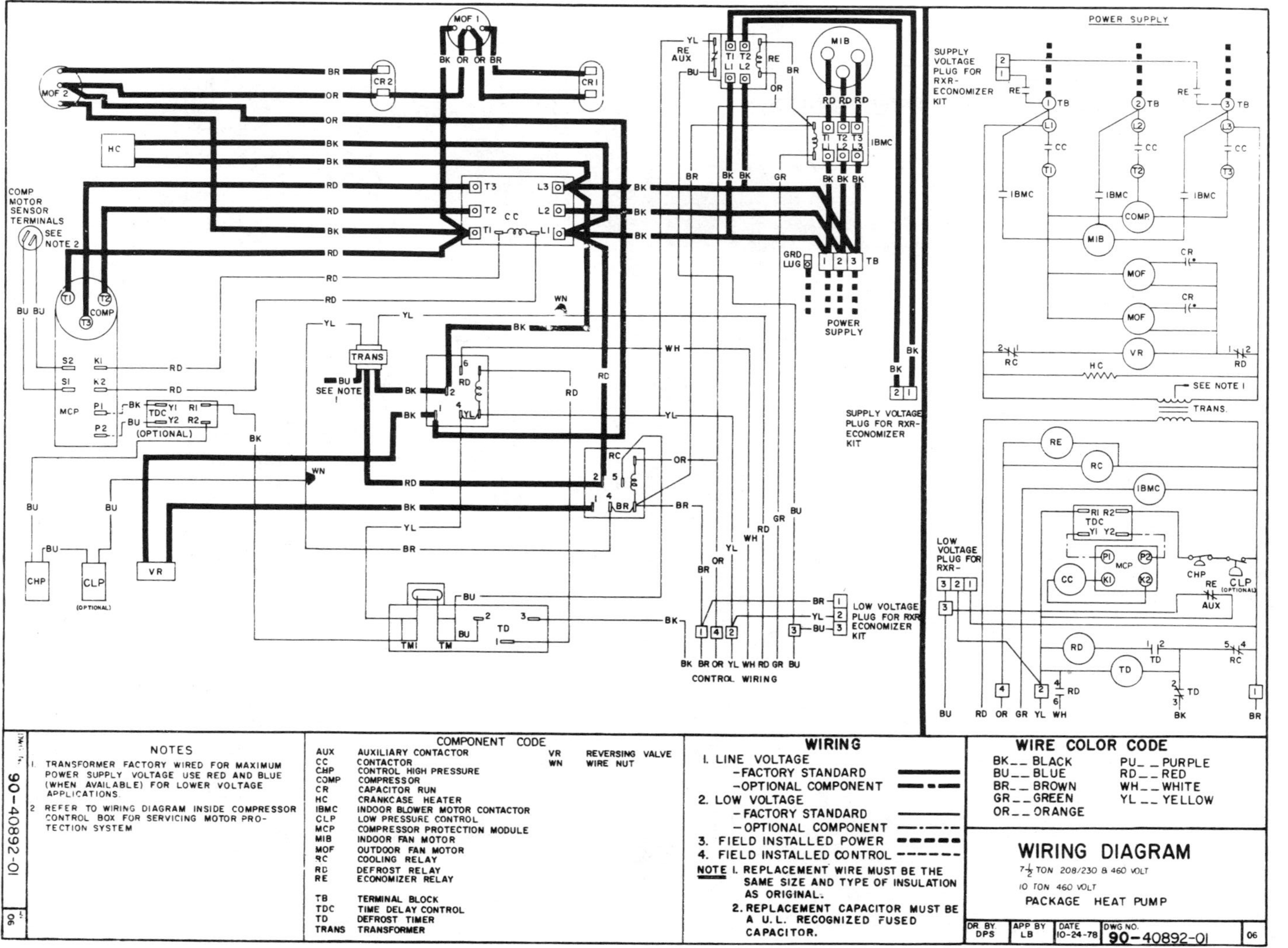
POWER SUPPLY
SUPPLY VOLTAGE PLUG FOR RXR-ECONOMIZER KIT
LOW VOLTAGE PLUG FOR RXR-ECONOMIZER KIT
SEE NOTE 1
TRANS.
CONTROL WIRING
COMP MOTOR SENSOR TERMINALS
SEE NOTE 2
NOTES
1. TRANSFORMER FACTORY WIRED FOR MAXIMUM POWER SUPPLY VOLTAGE USE RED AND BLUE (WHEN AVAILABLE) FOR LOWER VOLTAGE APPLICATIONS.
2. REFER TO WIRING DIAGRAM INSIDE COMPRESSOR CONTROL BOX FOR SERVICING MOTOR PROTECTION SYSTEM
COMPONENT CODE
AUX AUXILIARY CONTACTOR
CC CONTACTOR
CHP CONTROL HIGH PRESSURE
COMP COMPRESSOR
CR CAPACITOR RUN
HC CRANKCASE HEATER
IBMC INDOOR BLOWER MOTOR CONTACTOR
CLP LOW PRESSURE CONTROL
MCP COMPRESSOR PROTECTION MODULE
MIB INDOOR FAN MOTOR
MOF OUTDOOR FAN MOTOR
RC COOLING RELAY
RD DEFROST RELAY
RE ECONOMIZER RELAY
TB TERMINAL BLOCK
TDC TIME DELAY CONTROL
TD DEFROST TIMER
TRANS TRANSFORMER
VR REVERSING VALVE
WN WIRE NUT
WIRING
1. LINE VOLTAGE
-FACTORY STANDARD
-OPTIONAL COMPONENT
2. LOW VOLTAGE
-FACTORY STANDARD
-OPTIONAL COMPONENT
3. FIELD INSTALLED POWER
4. FIELD INSTALLED CONTROL
NOTE 1. REPLACEMENT WIRE MUST BE THE SAME SIZE AND TYPE OF INSULATION AS ORIGINAL.
2. REPLACEMENT CAPACITOR MUST BE A U.L. RECOGNIZED FUSED CAPACITOR.
WIRE COLOR CODE
BK__ BLACK
BU__ BLUE
BR__ BROWN
GR__ GREEN
OR__ ORANGE
PU__ PURPLE
RD__ RED
WH__ WHITE
YL__ YELLOW
WIRING DIAGRAM
7½ TON 208/230 & 460 VOLT
10 TON 460 VOLT
PACKAGE HEAT PUMP
DR BY DPS
APP BY LB
DATE 10-24-78
DWG NO. 90-40892-01
06
90-40892-01

TYPICAL WIRING SCHEMATIC

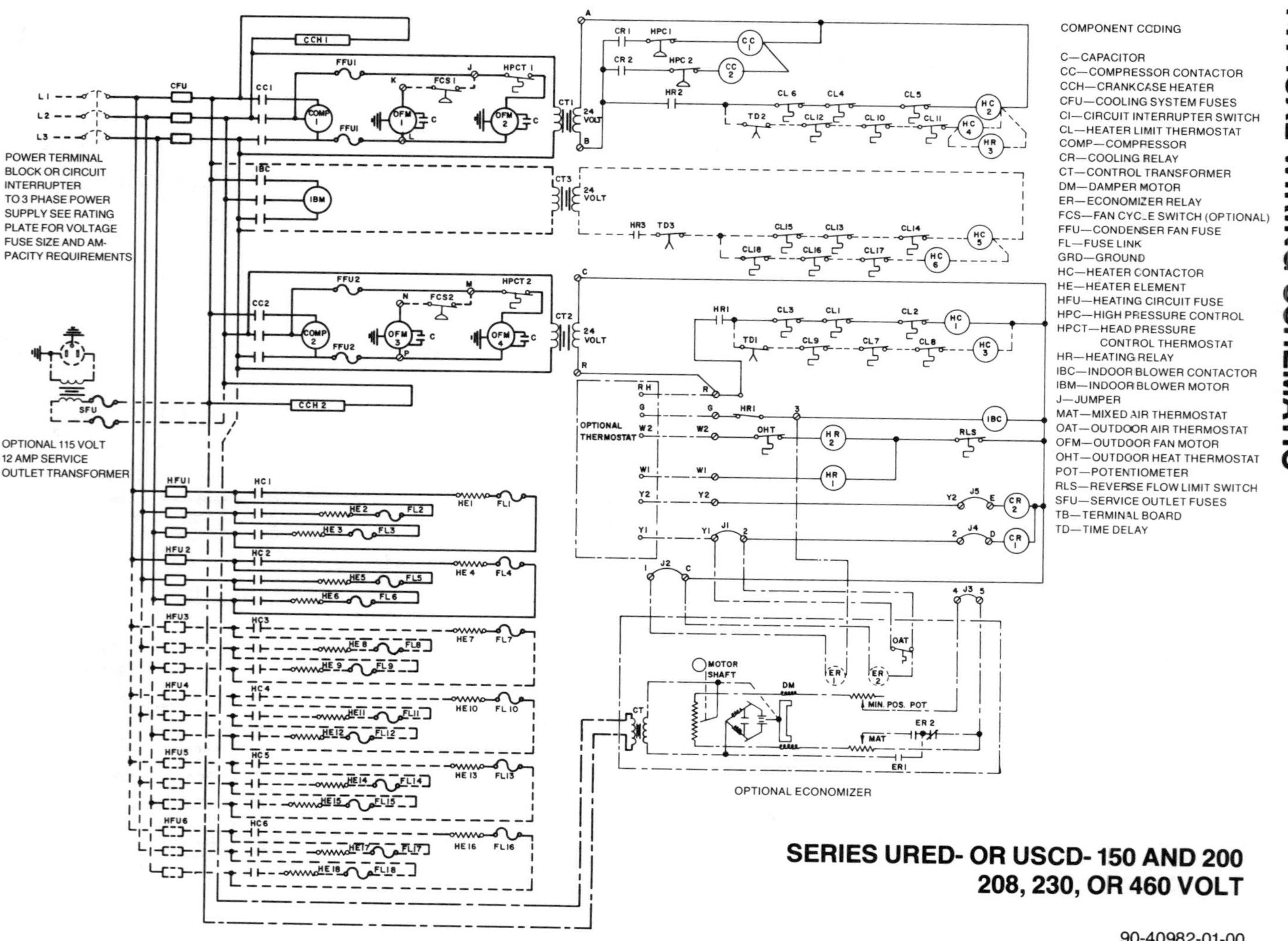

TYPICAL WIRING SCHEMATIC

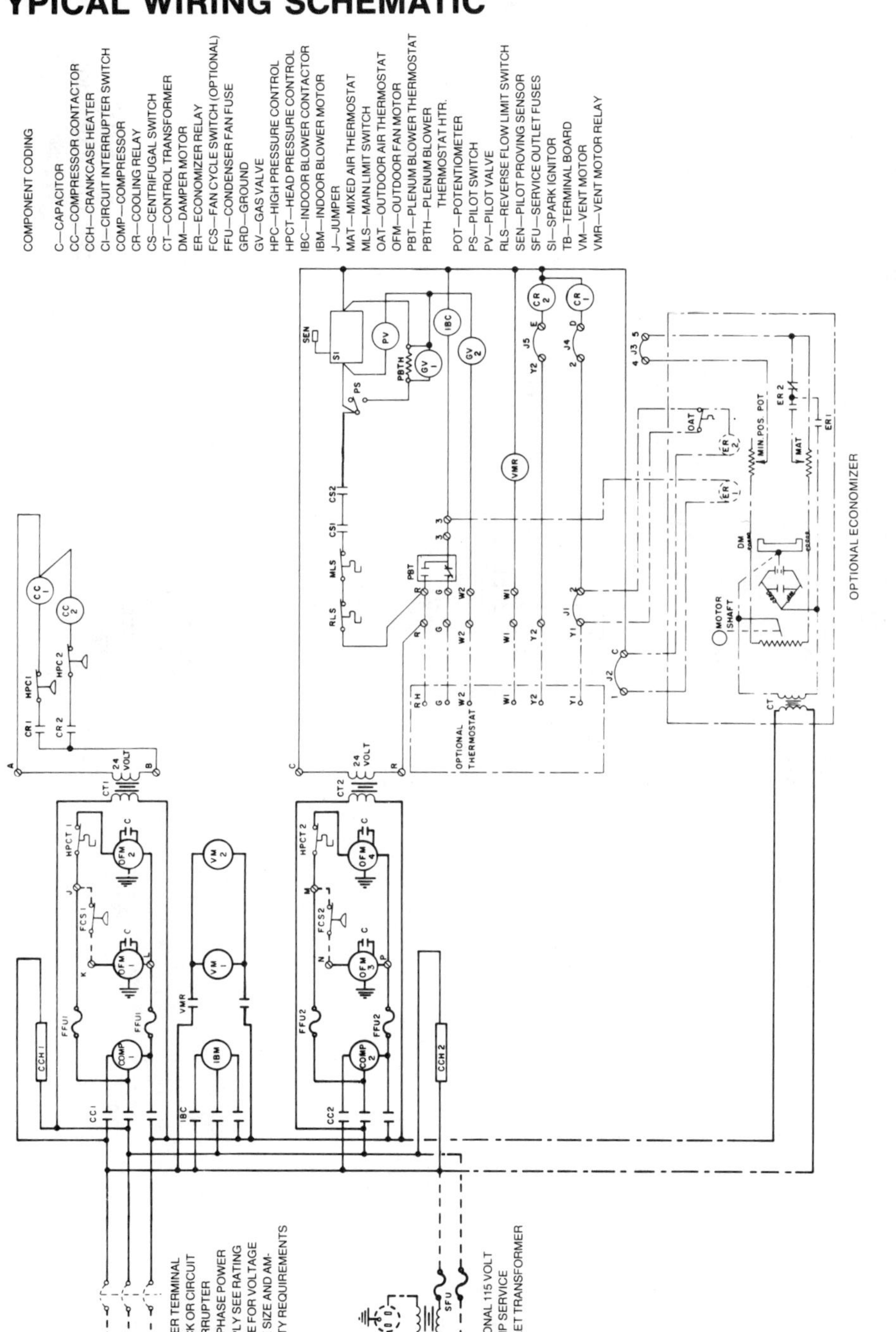

SERIES URGC-470-200
208, 230, OR 460 VOLT

90-41025-01-01

TYPICAL WIRING SCHEMATIC

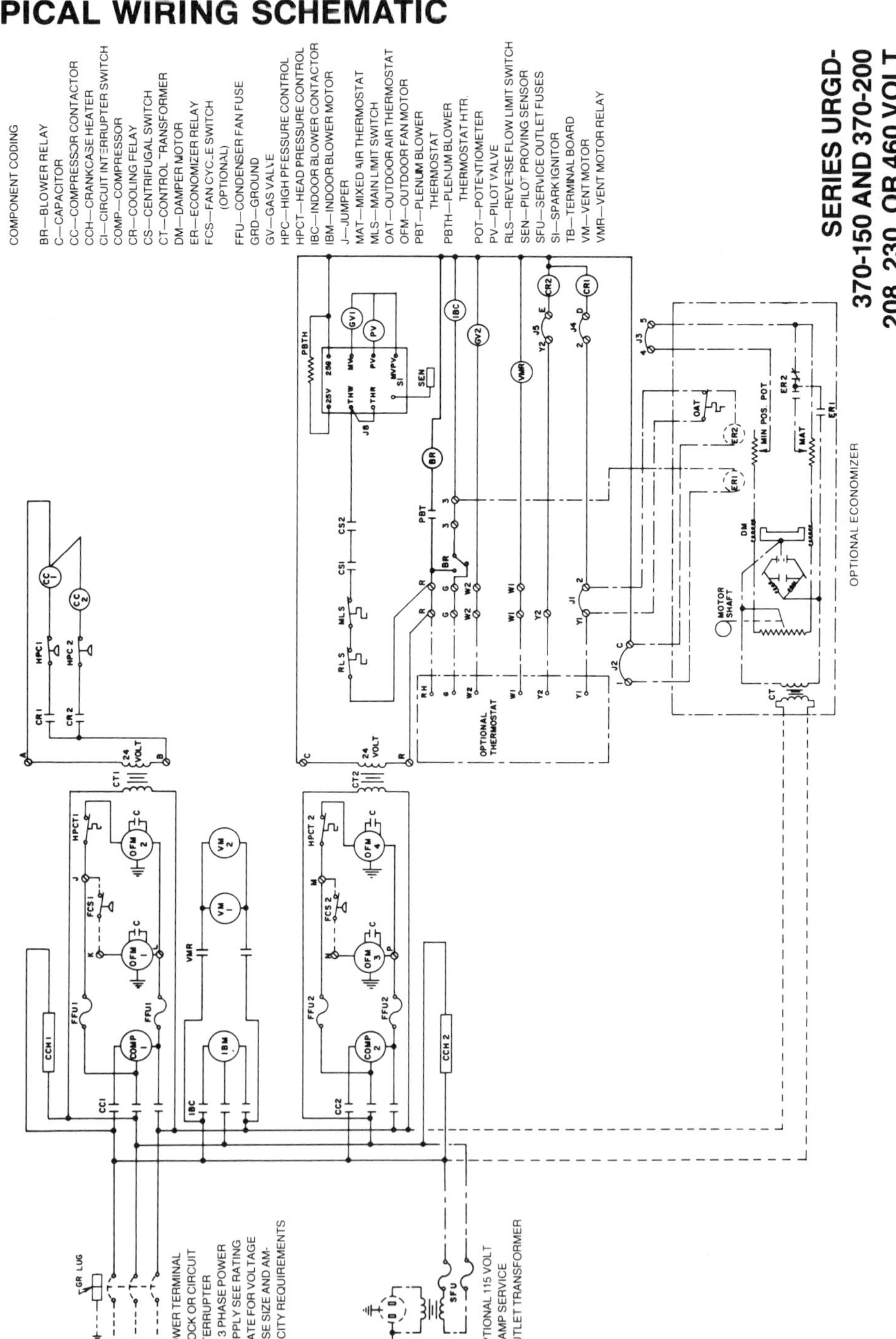

TYPICAL WIRING HOOK-UP DIAGRAM

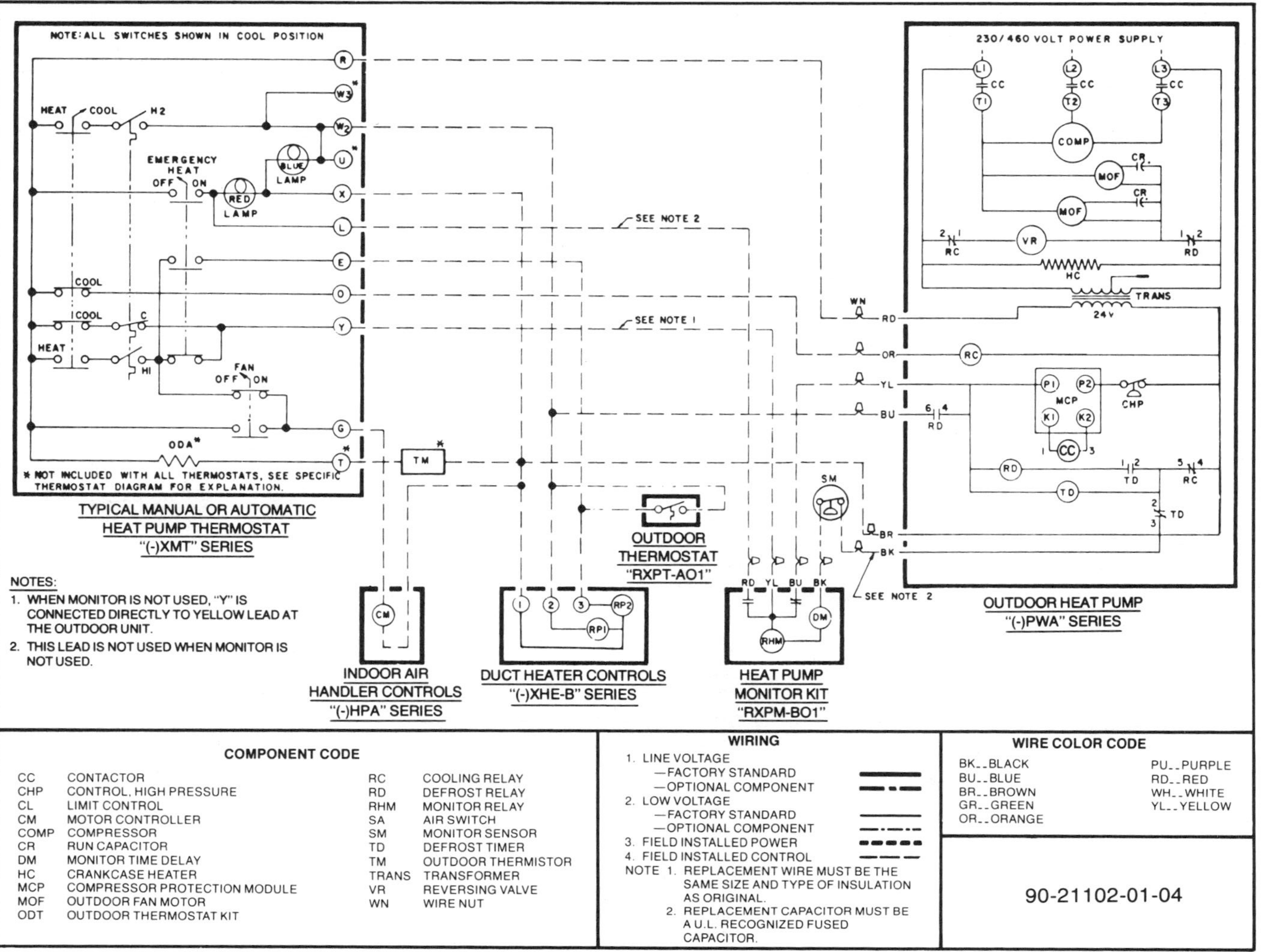

TYPICAL WIRING HOOK-UP DIAGRAM

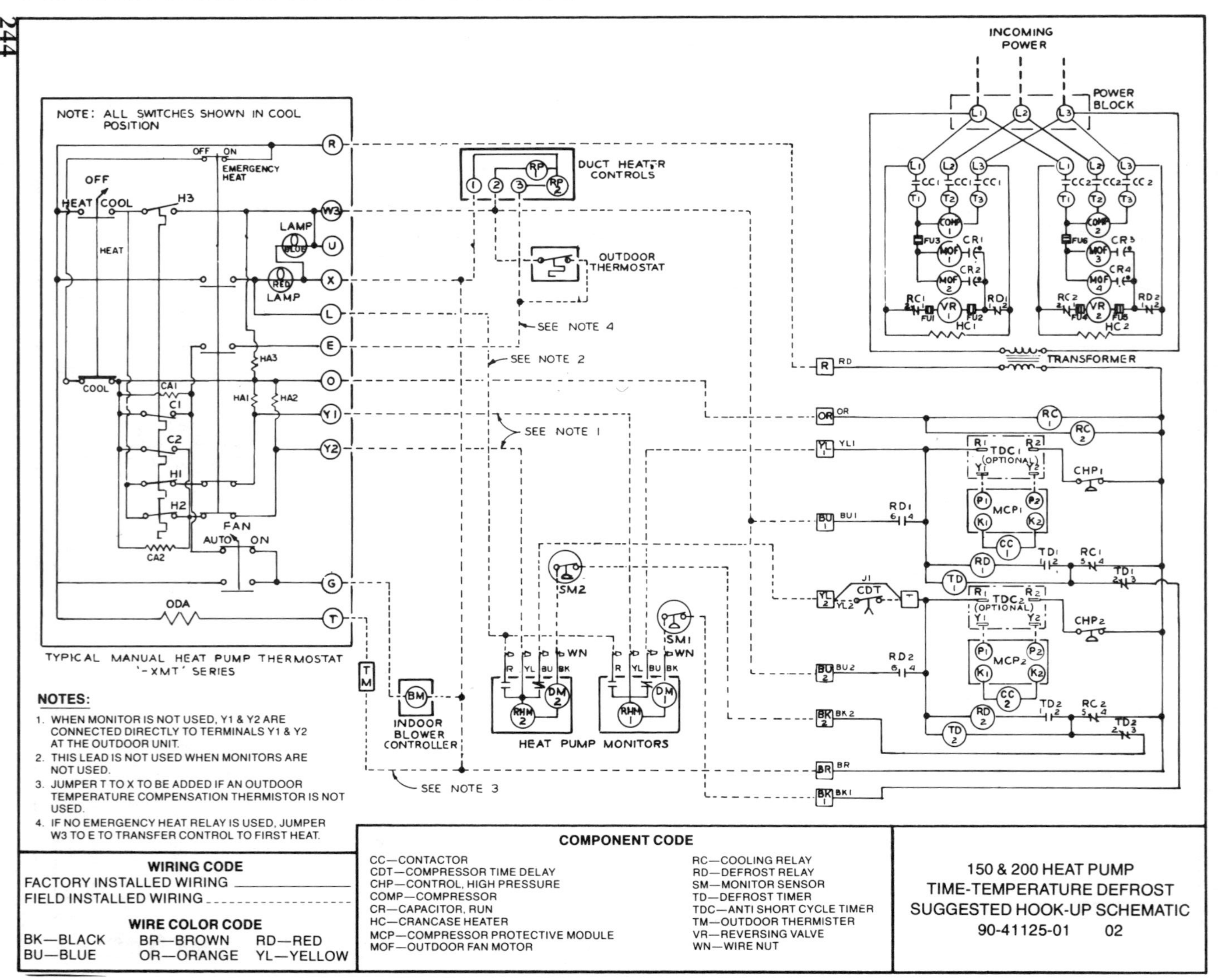

CHAPTER 6

Safety Procedures

The purpose of this list of safety procedures is to provide general safety precautions to be used by persons who own, install, operate, or maintain air conditioning and refrigeration equipment.

This list should not be used to replace instructions that are provided by the manufacturer. Therefore, anyone attempting to work on air conditioning and refrigeration equipment should be thoroughly familiar with the specific instructions for that particular unit.

The following is a list of the criteria that have been used here to indicate the intensity of the hazard:

1. **Danger** This means that there is an immediate hazard that will result in severe personal injury or death.
2. **Warning** This means that hazards or unsafe practices could result in severe injury or death.
3. **Caution** This means that potential hazards or unsafe practices could result in minor personal injury.
4. **Safety Instructions** These are general instructions that are necessary for safe working practices.

Personal Protection

Warning

1. Do not touch electrical wiring connections with wet hands.
2. Do not touch electrical equipment while standing on a wet surface or wearing wet shoes.
3. Do wear a hard hat or other head protection when there is a possibility of falling objects.

Caution

1. Do wear safety glasses equipped with side shields when working in manufacturing plants or at construction sites.
2. Do wear gloves when handling system components after a compressor motor burnout. The refrigerant and oil contain acid that can result in acid burns to the skin.
3. Do wear goggles and gloves when handling chemicals; when welding, cutting, grinding, or brazing; or when in an area where these operations are performed.
4. Do wear gloves and other protective clothing when working with sheet metal.
5. Do wear safety shoes when working around or lifting heavy objects.
6. Do wear protective clothing when arc welding to protect from serious burns.
7. Do wear hearing protection when working in areas where sound levels are greater than 90 db.
8. Do not wear rings, jewelry, loose clothing, long ties, or gloves while working around moving belts and machinery.
9. Do not wear rings or watch while working on electrical equipment.

Safety Instructions

1. Keep your work area clean of debris and free of liquid spills on the floor.
2. Do not continue working if you become seriously ill. An ill person is less observant and is therefore more subject to accidents.

Rigging (Use of Cranes)

Danger

1. Never use cranes under power lines. They may come in contact with the lines, causing a high voltage electrical short.

Warning

1. Do check for the center of gravity before hoisting heavy equipment.

2. Do check for any specific hoisting instructions by the manufacturer before hoisting equipment.
3. Do check the component and assembly weights before assembling equipment to make certain the crane can lift the unit safely.
4. Do use only approved methods and rigging equipment.
5. Do use eyebolt holes to hoist an entire assembly.
6. Do not move a loaded hoist, crane, or chain fall until the path is clear.
7. Do not use faulty rigging equipment.

Caution

1. Do not use rigging equipment when there is a possibility of slipping or losing your balance.
2. Do use platforms or catwalks to cross over a machine. Never climb over a machine.
3. Do not use ladders that are too straight up or have too much slope from top to bottom.

Safety Instructions

1. Do use lifting lugs according to hoisting instructions.
2. Do keep aware of where fellow workmen are at all times when hoisting equipment.
3. Do post signs indicating that heavy objects are being hoisted.

Storing and Handling Refrigerant Cylinders

Warning

1. Do not heat a refrigerant cylinder with an open flame. When necessary to heat a cylinder, use warm water.
2. Do not store refrigerant cylinders in direct sunlight.
3. Do not store refrigerant cylinders where the surrounding temperature may exceed the relief valve setting.
4. Do not reuse disposable refrigerant cylinders. It is dangerous and illegal.
5. Do not attempt to burn or incinerate refrigerant cylinders.

6. Do not alter the safety devices on a refrigerant cylinder.
7. Do not force connections.
8. Do not overfill refillable refrigerant cylinders.
9. Do open cylinder valves slowly to prevent rapid overpressurizing of the system.
10. Do use a proper wrench when opening and closing refrigerant cylinder valves.

Caution

1. Do not alter refrigerant cylinders.
2. Do not drop, dent, or abuse refrigerant cylinders.
3. Do not charge a cylinder with refrigerant different from the color coding.
4. Do not entirely depend on the color code of a refrigerant cylinder for refrigerant identification.
5. Do not overcharge a rechargeable refrigerant cylinder.
6. Do replace cylinder caps when cylinders are not in use.
7. Do avoid pressure surges when transferring refrigerant from one cylinder to another.
8. Do periodically inspect all hoses, fittings, and charging manifolds and replace them when needed.
9. Do secure all refrigerant cylinders to prevent damage. The following procedure is recommended.
    a. Large cylinder: Lay on its side and prevent rolling by using chocks.
    b. Small cylinder: Store in an upright position and secure with a strap or chain.

Leak Testing and Pressure Testing Systems

Danger

1. Do not use oxygen for pressurizing a refrigeration system. Oxygen and oil combine to cause an explosion.
2. Do not use full cylinder pressure when pressurizing a system with nitrogen.

3. Do not exceed the specified system test pressures when pressurizing a system.
4. Do not pressurize a system with nitrogen before putting in the refrigerant when leak-testing. The nitrogen may have a greater pressure than the refrigerant cylinder can withstand.
5. Do use nitrogen when pressurizing a system above refrigerant pressures.
6. Do use a gauge-equipped regulator when pressurizing a system with nitrogen.
7. Do disconnect the nitrogen cylinder from the system when the system is pressurized.

Refrigerants

Warning

1. Do not enter an enclosed area after a refrigerant leak has occurred without thoroughly ventilating the space. Use the buddy system or an approved air pack or both.
2. Do not allow liquid refrigerant to come into contact with the skin or eyes. Immediately wash the skin with soap and water. Immediately flush the eyes with water and consult a physician.
3. Do not breathe fumes given off by a leak detector or open flame. The fumes are likely to be phosgene, a deadly gas.
4. Do use safety goggles.
5. Do use gloves when working with liquid refrigerants.

Caution

1. Do not weld or cut a refrigerant line or vessel until all the refrigerant has been removed.
2. Do not use an open flame in a space containing refrigerant vapor. Properly ventilate the area before entering.
3. Do not smoke in a space filled with refrigerant vapor.

Safety Instructions

1. Avoid heavy concentrations of refrigerant within an enclosed area. A refrigerant can displace enough oxygen to cause suffocation.

2. Do not allow heating devices, such as gas flames or electric elements, to operate in an area filled with refrigerant vapor. Heat can decompose the refrigerant into hazardous substances such as hydrochloric acid, hydrofluoric acid, and phosgene gas.
3. If a strong, irritating odor is detected, warn other persons and immediately leave the area. Report the problem to the proper persons.

Reciprocating Compressors

Danger

1. Do not work on electrical wiring until all electrical power is off.
2. Do not take ohmmeter measurements with the electrical power on.

Warning

1. Do not use a hermetic compressor for a vacuum pump. The winding can short out or a terminal may blow out, causing serious injury.
2. Do not use a welding torch when removing a compressor from the refrigerant system. The oil could catch fire and cause serious burns.
3. Do not purge refrigerant from the system through a cut, loosened, or broken pipe. Use the gauge manifold so that the rate of purging can be controlled.
4. Do not apply voltage to a compressor motor while the terminal box cover is off.
5. Do not attempt to loosen or remove bolts from the compressor while it is under pressure. Pump the system down to 0 to 2 psig (100.989 kPa to 114.73 kPa).
6. Do not attempt to operate a compressor with the suction and discharge service valves closed.
7. Do open, tag, and lock all electrical disconnect switches while servicing the electrical circuits and connections.

Caution

1. Do valve off all the compressors in a multiple-compressor system before attempting to service any circuit. Otherwise, the oil equalizer connection will prevent purging pressure from a single compressor.

Air Handling Equipment

Danger

1. Do not enter an enclosed fan cabinet while the unit is running.
2. Do not reach into a unit or fan cabinet while the unit is running.
3. Do not work on a fan or fan motor until the electrical disconnect switch is open, tagged, locked, and the fuses removed.
4. Do not work on electric circuits, heating elements, or connections until the electrical disconnect switch is open, tagged, locked, and the fuses removed.

Warning

1. Do not operate belt-driven equipment without the belt guards in place.
2. Do not service air control dampers until the operators are disconnected.
3. Do not handle access covers in high winds without sufficient help.
4. Do not pressurize a coil with a liquid refrigerant for leak testing.
5. Do not steam-clean coils until all personnel are clean of the area.
6. Do ensure that there is adequate ventilation when welding or cutting inside an air handling unit.
7. Do be sure that rooftop units are properly grounded.

Caution

1. Do not work on fans without securing the sheave with a rope or strap to prevent fan freewheeling.
2. Do not exceed the specified test pressure when pressurizing the system.
3. Do protect all flammable material when welding or cutting inside an air handling unit.

Safety Instructions

1. When operating and servicing air handling equipment, use good

judgment and safe working practices to prevent damage to equipment and personal injury or property damage.

Oxyacetylene Welding and Cutting

Danger

1. Do not use oxygen for any purpose except welding and cutting.
2. Do not use oxygen to pressurize refrigeration systems.

Warning

1. Do not store oxygen cylinders near oil and/or grease.
2. Do not store oxygen cylinders near combustible material.
3. Do not use oily hands or gloves to handle oxygen cylinders.
4. Do not weld or cut in an atmosphere filled with refrigerant vapor.
5. Do not weld or cut near combustible materials.
6. Do not weld or cut lines or pressure vessels until they have been properly evacuated.
7. Do not weld or cut unless adequate ventilation is provided.
8. Do use an approved breathing system and a buddy system when necessary to weld or cut in an unventilated area.
9. Do wear goggles and welding gloves when welding or cutting.

Caution

1. Do not store oxygen and acetylene cylinders next to each other.
2. Do not store oxygen and acetylene cylinders near a heat source.
3. Do not block passageways, stairways, or ladders with welding equipment.
4. Do store oxygen and acetylene cylinders strapped or chained in an upright position.
5. Do wear suitable protective clothing when welding or cutting.

Safety Instructions

1. Do not use damaged or worn hoses.

2. Do not use connectors other than those made specifically for oxyacetylene welding and cutting equipment.
3. Do not stand in front of the regulator when opening a cylinder valve.
4. Do not ignite torches with anything other than a friction lighter.
5. Do observe the color coding of pipelines, cylinders, and hoses.
6. Do crack the cylinder valve before attaching the pressure regulator.
7. Do release tension on the regular adjusting screw before opening the cylinder valve.
8. Do inspect the equipment for leaking shut-off valves and connections before use.

Refrigeration and Air Conditioning Machinery (General)

Warning

1. Do not attempt to siphon refrigerants or other chemicals by mouth.
2. Do not attempt to weld, cut, or remove fittings while the system is pressurized.

Caution

1. Do not attempt to bend or step on pressurized refrigerant lines.
2. Do not weld or cut in an area containing refrigerant.
3. Do not loosen a packing gland nut before making sure there are plenty of threads engaged to prevent a blowout.
4. Do use replacement parts that meet the requirements of the equipment.
5. Do tag and valve off refrigerant, water, and steam lines before opening them.
6. Do periodically inspect fittings, piping, and valves for corrosion, leaks, damage, or rust.

Safety Instructions

1. Do be sure that all shipping bolts and plugs have been removed before initial start-up.
2. Do periodically check all refrigerant and oil sight glasses for cracks.

3. Do use an ice-melting spray to remove ice from sight glasses.
4. Do not chip ice from sight glasses.

Centrifugal Liquid Chillers (Heat Exchangers)

Danger

1. Do not exceed the specified test pressures when pressurizing a system.
2. Do not use oxygen to pressurize, purge, or leak-test a refrigeration system.
3. Do not make any pressure-relieving device inoperative.
4. Do not operate any machine without proper pressure relief devices installed and functioning.

Warning

1. Do not attempt to repair any pressure relief device.
2. Do not make any pressure-relieving device inoperative.
3. Do not vent pressure-relieving devices inside a building.
4. Do not open a refrigerant system that is under pressure.
5. Do inspect all pressure relief devices at least once a year.
6. Do check the type of refrigerant before charging the system.
7. Do properly ventilate the area that contains refrigerant vapor before attempting to make repairs.

Caution

1. Do not loosen water box cover bolts without completely draining the water box.
2. Do not install shut-off valves on both sides of a component in such a way that liquid might be trapped.
3. Do install a drain in the vent line near each pressure relief device.

Centrifugal Liquid Chillers (Electrical Circuits and Controls)

Danger

1. Do not attempt to check voltage without proper instructions.

2. Do not attempt to take high voltage measurements (600 volts or more) with hand-held instrucments.
3. Do use current and potential transformers when making high-voltage measurements.

Warning

1. Do not work on electrical circuits until they are opened, tagged, and locked.
2. Do not attempt to stop a unit by opening a knife switch.
3. Do not attempt to check the resistance on an energized circuit.
4. Do not remove terminal box covers while the compressor is running.
5. Do not handle capacitors until they have been properly discharged.
6. Do ground all electrical equipment.
7. Do use a ground-fault current interruptor when using hand tools.

Caution

1. Do not bypass or block electrical interlocks.
2. Do not check an electrical circuit until you are certain that the electric power is off in all circuits.
3. Do use insulated fuse pullers on cartridge-type fuses.

Centrifugal Liquid Chillers (Couplings)

Danger

1. Do not remove guards until all rotating parts have completely stopped.
2. Do not start a unit until all bolts are tight.
3. Do be sure that all tools and materials have been removed before starting a unit.

Warning

1. Do not stand beside rotating couplings.
2. Do not operate a beltdrive unit without the guards in place.

Caution

1. Do not start a unit until all wrenches, dial indicators, etc., have been removed.
2. Do tighten all coupling bolts twice to make certain that they are tight.

Centrifugal Chillers (Turbines)

Danger

1. Do not start a unit until all superheated steam connections have been checked.
2. Do not stand in line with turbine trip valve handles.
3. Do familiarize yourself with the proper start-up and operation procedures before starting the unit.

Warning

1. Do not open drain lines on a turbine while the unit is under a vacuum.
2. Do not block open a steam governor.
3. Do not block a steam trip handle.
4. Do not block a pressure relief valve.
5. Do not operate turbines at speeds greater than their designed speed.
6. Do install a relief valve between the turbine exhaust flange and the first shut-off valve to protect against rupturing the turbine casing.
7. Do install the proper blinds or blank flanges in the inlet and exhaust lines before dismantling a turbine.

Caution

1. Do not touch a hot turbine or components without the proper protective clothing.
2. Do keep the area clean of oil spills and oily rags.

Safety Instructions

1. Do not attempt to operate a turbine until all preoperative safety devices and control devices have been checked.
2. Do check and adjust turbine safeties and speed trip devices annually.

Absorption Liquid Chillers

Danger

1. Do not exceed the specified test pressures when pressurizing a refrigerant system.
2. Do not use oxygen to purge lines or to pressurize a refrigeration system.

Warning

1. Do not attempt to remove fittings or components while the system is under pressure.
2. Do not siphon lithium bromide by mouth.
3. Do not flame-cut a purge chamber until all the hydrogen has been evacuated.
4. Do not service an electrical circuit until the switch is opened, tagged, and locked.
5. Do wear goggles and the proper protective clothing when handling inhibitor, octyl alcohol, lithium hydroxide, hydrobromic acid, and lithium bromide.
6. Do immediately wash any chemical spills from the skin with soap and water.
7. Do immediately flush your eyes with water and consult a physician if your eyes have been exposed to chemicals.
8. Do properly ventilate the area when welding or cutting to remove any noxious fumes.

Caution

1. Do not loosen cover bolts before draining water boxes.
2. Do keep the floor clean of spills and debris.
3. Do tag and valve off water, steam, and brine lines before opening them.

INDEX